Recent Titles in This Series

(Continued in the back of this publication)

MEMOIRS
of the
American Mathematical Society

Number 493

Lattice Structures
on Banach Spaces

Nigel J. Kalton

May 1993 • Volume 103 • Number 493 (end of volume) • ISSN 0065-9266

American Mathematical Society
Providence, Rhode Island

1991 *Mathematics Subject Classification.*
Primary 46B30, 46E30; Secondary 46E05.

Library of Congress Cataloging-in-Publication Data

Kalton, Nigel J. (Nigel John), 1946–
  Lattice structures on Banach spaces/Nigel J. Kalton.
  p. cm. – (Memoirs of the American Mathematical Society, ISSN 0065-9266; no. 493)
  "Volume 103, number 493 (end of volume)."
  Includes bibliographical references.
  ISBN 0-8218-2557-7
  1. Banach lattices. I. Title. II. Series.
QA3.A57 no. 493
[QA326]
510 s–dc20   93-466
[512′.55]   CIP

Memoirs of the American Mathematical Society

This journal is devoted entirely to research in pure and applied mathematics.

Subscription information. The 1993 subscription begins with Number 482 and consists of six mailings, each containing one or more numbers. Subscription prices for 1993 are $336 list, $269 institutional member. A late charge of 10% of the subscription price will be imposed on orders received from nonmembers after January 1 of the subscription year. Subscribers outside the United States and India must pay a postage surcharge of $25; subscribers in India must pay a postage surcharge of $43. Expedited delivery to destinations in North America $30; elsewhere $92. Each number may be ordered separately; *please specify number* when ordering an individual number. For prices and titles of recently released numbers, see the New Publications sections of the *Notices of the American Mathematical Society.*

Back number information. For back issues see the *AMS Catalog of Publications.*

Subscriptions and orders should be addressed to the American Mathematical Society, P. O. Box 1571, Annex Station, Providence, RI 02901-1571. *All orders must be accompanied by payment.* Other correspondence should be addressed to Box 6248, Providence, RI 02940-6248.

Copying and reprinting. Individual readers of this publication, and nonprofit libraries acting for them, are permitted to make fair use of the material, such as to copy a chapter for use in teaching or research. Permission is granted to quote brief passages from this publication in reviews, provided the customary acknowledgement of the source is given.

Republication, systematic copying, or multiple reproduction of any material in this publication (including abstracts) is permitted only under license from the American Mathematical Society. Requests for such permission should be addressed to the Manager of Editorial Services, American Mathematical Society, P. O. Box 6248, Providence, RI 02940-6248.

The owner consents to copying beyond that permitted by Sections 107 or 108 of the U.S. Copyright Law, provided that a fee of $1.00 plus $.25 per page for each copy be paid directly to the Copyright Clearance Center, Inc., 27 Congress Street, Salem, MA 01970. When paying this fee please use the code 0065-9266/93 to refer to this publication. This consent does not extend to other kinds of copying, such as copying for general distribution, for advertising or promotion purposes, for creating new collective works, or for resale.

Memoirs of the American Mathematical Society is published bimonthly (each volume consisting usually of more than one number) by the American Mathematical Society at 201 Charles Street, Providence, RI 02904-2213. Second-class postage paid at Providence, Rhode Island. Postmaster: Send address changes to Memoirs, American Mathematical Society, P. O. Box 6248, Providence, RI 02940-6248.

TABLE OF CONTENTS

v

ABSTRACT

In the memoir [22], Johnson, Maurey, Schechtmann and Tzafriri initiated the study of the possible rearrangement-invariant lattice structures on an arbitrary Banach space. In this work, using somewhat different techniques, we continue this study and improve their main results on uniqueness of structure as an r.i. function space on $[0, 1]$. Our techniques also yield more general results on uniqueness of structure as a Banach lattice. For example, if X is an order-continuous r.i. space on $[0, 1]$ and $X \neq L_2$ then if X embeds as a complemented subspace of an order-continuous Banach lattice Y then X embeds as a complemented sublattice; furthermore this sublattice is lattice-isomorphic either to X or, if the Haar basis of X is unconditional, to the atomic Banach lattice generated by the Haar basis. We also show that if Y is a nonatomic Banach lattice which is q-concave for some $q < 2$ and is isomorphic to an r.i. function space X on $[0, 1]$ then Y is lattice-isomorphic to X. This leads to examples of r.i. spaces with unique structure as nonatomic Banach lattices (in addition to the known examples L_1 and L_2 due to Abramovich and Wojtaszczyk).

Key words and phrases: Banach lattice, uniqueness of lattice structure, non-atomic, Köthe function space.

1. Introduction

The general problem we address in this paper is to attempt to characterize in some way the possible Banach lattice structures that a separable Banach space may have. Thus given two Banach lattices X and Y which are isomorphic as Banach spaces we attempt to derive results concerning their respective lattice structures; in extreme cases we may hope for results that assert that X and Y are lattice-isomorphic. In this paper we will be mainly concerned with the case of order-continuous nonatomic Banach lattices which therefore may be concretely represented as Köthe function spaces. However we review first the literature for general lattices.

Of course a separable order-continuous atomic Banach lattice is nothing other than a Banach space with a given unconditional basis. The question of uniqueness of unconditional basis was considered first by Lindenstrauss, Pelczynski and Zippin in the late sixties ([30],[32],[34]). In the language of this paper they showed that there are exactly three spaces (ℓ_1, ℓ_2 and c_0) with a symmetric unconditional basis which have exactly one structure as an atomic order-continuous Banach lattice. Subsequently, Edelstein and Wojtaszczyk ([18],[48],[49]) showed that direct sums of these spaces such as $\ell_1 \oplus \ell_2$ also have unique structure as atomic order-continuous lattices. In the memoir [5], Bourgain, Casazza, Lindenstrauss and Tzafriri probed the problem of characterizing spaces with such a unique structure and gave further examples such as $\ell_1(\ell_2)$. In a different direction the uniqueness of the symmetric basis was investigated in [31] and [42].

The first result on uniqueness of structure for nonatomic Banach lattices seems to be that of Abramovich and Wojtaszczyk [2] who showed that the the uniqueness of the unconditional basis for ℓ_1 and ℓ_2 has an analogue for the function spaces L_1 and L_2. These spaces have unique structure as nonatomic Banach lattices.

The general study of corresponding problems for nonatomic Banach lattices was, however, initiated in the seminal work of Johnson, Maurey, Schechtman and Tzafriri [22], and

Received by the editor April 3, 1991 and in revised form, November 22, 1991

This research was supported by NSF-grants DMS-8601401 and DMS-8901626

it is this paper which provides the motivation for the present work. In [22], the study was focussed on rearrangement-invariant (r.i.) spaces on either $[0, 1]$ or $[0, \infty)$. These spaces are, in a certain sense, the continuous analogues of spaces with symmetric bases. The main aim was to prove results characterizing the possible r.i. structures on a given Banach space. We will discuss their main results in more detail in a moment. First, however, we also mention that their work also initiated the study of finite-dimensional symmetric spaces and the general problem of uniqueness of symmetric basis in this case (see [6], [13], [20] and [44]). We also mention that in [13] and [44] these ideas were extended to the study of possible lattice structures on finite-dimensional Banach spaces, with results which when stated in the appropriate language resemble some of the theorems in this paper in the infinite-dimensional setting; typical results asserted that under suitable hypotheses, if two finite-dimensional Banach lattices are isomorphic then they contain large dimensional bands which are lattice-isomorphic (cf. Theorem 2.6 of [13] or Theorems 5.4 and 5.5 of [44]).

It is well-known that a separable r.i. space on $[0, 1]$ has an unconditional basis and hence a structure as an atomic Banach lattice if and only its Boyd indices satisfy $1 < p_X \leq q_X < \infty$, and in this case the Haar system gives an unconditional basis of X (see [22], [33]). We shall refer to the induced atomic Banach lattice as the Haar representation of X, denoted by H_X.

In [22] the approach taken to uniqueness of structure theorems is via embedding theorems. Thus, given an r.i. space Y on $[0, 1]$, one tries first to identify those r.i. spaces X which can be embedded into Y. One way to embed X into Y (when $1 < p_X \leq q_X < \infty$) is to embed H_X as a sublattice; the conclusions of Theorems 5.1 and 6.1 of [22] are that, if we exclude this possibility then, under suitable hypotheses, the identity map is bounded from X to Y. Precisely:

THEOREM 1.1 [22]. *Let X and Y be separable r.i. spaces on $[0, 1]$ and assume that Y is q-concave for some $q < \infty$. Suppose X is isomorphic to a subspace of Y. Suppose either:*
(a) $1 < p_X \leq q_X < \infty$, X is not isomorphic as an r.i. space to L_2, and H_X is not lattice-isomorphic to a sublattice of Y, or:
(b) $p_X = 1$ and for some $r < 2$ there exists C such that $\|f\|_X \leq C\|f\|_r$ for all $f \in X$.
Then there is a constant C_0 so that for all $f \in X$ we have $\|f\|_Y \leq C_0\|f\|_X$.

From this theorem results on the uniqueness of r.i. structure can be obtained. Thus if we have that X is isomorphic to Y we utilize the facts that X embeds into Y and Y embeds into X; if we want to exploit the fact that X is isomorphic to a complemented subspace of Y we use a duality argument that X embeds into Y and X^* embeds into Y^*. Thus typical uniqueness of structure results obtained are:

THEOREM 1.2 [22], [33]. *Let X and Y be separable r.i. function spaces on $[0,1]$ and suppose X and Y are isomorphic. Then each of the following hypotheses imply that $X = Y$ as an r.i. space:*
(a) X is r-concave for some $r < 2$.
(b) X is superreflexive and H_X is not lattice-isomorphic to a sublattice of X.

Subsequent work on r.i. spaces has concentrated on characterizing subspaces of r.i. spaces (see [7], [8], [9], [10], [11], [15], [16], [23], [40], [41], [45]). In this paper we concentrate on the problems of identifying complemented subspaces (see [12] and [21]) and the question of uniqueness of lattice structure. In the course of our study we give a more complete result on uniqueness of r.i. structure, eliminating concavity hypotheses and strengthening the conclusions.

By way of motivation let us consider two situations. First suppose X is a reflexive Banach space with two unconditional bases (x_n) and (y_n). Then by a standard argument it is clear that there is a block basic sequence (z_n) of (y_n) equivalent to a subsequence (x_{k_n}) of (x_n) and spanning a complemented subspace of X. This can be rephrased: if X and Y are separable reflexive atomic Banach lattices which are isomorphic then there is an infinite-dimensional band B in X which is lattice-isomorphic to a complemented sublattice of Y. The second example is the result of [25] which shows that something similar occurs for a special nonatomic Banach lattice: if X is a Banach lattice not containing c_0 and L_1 embeds into X then L_1 embeds as a complemented sublattice of X.

We prove a very general result of this type in Theorem 7.2. We suppose that Y is a separable order-continuous Banach lattice which contains no complemented sublattice lattice-isomorphic to ℓ_2; we suppose further that X is a separable order-continuous Banach lattice which contains no complemented sublattice lattice-isomorphic to L_2. Then if X is isomorphic to a complemented subspace of Y there is a nontrivial band X_0 in X which is lattice-isomorphic to a complemented sublattice of Y. More generally X itself is lattice-isomorphic to a complemented sublattice of $Y(\ell_2)$. This latter result is the nonatomic generalization of a theorem due to Maurey [37] (cf. [33]) on unconditional bases in Banach lattices.

In the case when we suppose that X is an r.i. space these results can be sharpened. In Theorem 7.3 we will show that if Y is a separable order-continuous Banach lattice and if X is a separable r.i. space on $[0,1]$ with $X \neq L_2$ and if X is isomorphic to a complemented subspace of Y then either $1 < p_X \leq q_X < \infty$ and H_X is lattice-isomorphic to a complemented sublattice of Y or X itself is lattice-isomorphic to a complemented sublattice of Y. Thus if a separable r.i. space on $[0,1]$ is complementably embeddable in a separable order-continuous Banach lattice it is embeddable as a complemented sublattice. This result is apparently new for the case L_p when $p > 1$; for $p = 1$ it is true without the hypothesis of complementation (see [25] and Theorem 10.7, below).

These results allow an improvement of Theorem 1.2 above. We show that if X and Y are isomorphic separable r.i. spaces on $[0,1]$ such that either (a) $1 < p_X \leq q_X < \infty$ and H_X is not lattice-isomorphic to a complemented sublattice of X or (b) $p_X = 1$ or (c) $q_X = \infty$ then $X = Y$ as an r.i. space. This improves on the theorems of [22] by removing all restrictive hypotheses on X and by requiring in the excluded case that H_X be a *complemented* sublattice.

There is obviously no hope except in special cases of establishing uniqueness of the nonatomic lattice structure for an r.i. space. Indeed it is shown in [22] that if $1 < p_X \leq q_X < \infty$ then X admits many nonatomic lattice structures. For example X is isomorphic to $X(\ell_2)$, $X(L_2)$ and to an r.i. space on $[0,\infty)$. However in the special case of L_p they establish a uniqueness theorem by eliminating structures in which ℓ_2^n can be embedded uniformly as a (complemented) sublattice. In the terminology we introduce in the paper this is equivalent to showing that if $1 < p < 2$ then L_p has a unique structure as a strictly 2-concave (i.e. r-concave for some $r < 2$) Banach lattice and if $2 < p < \infty$ then L_p has unique structure as a strictly 2-convex (i.e. r-convex for some $r > 2$) Banach lattice. This result is a corollary of a stronger result that any strictly 2-convex nonatomic Banach lattice which is isomorphic to a subspace of L_p for $p > 2$ is lattice-isomorphic to L_p. The atomic version of these results for ℓ_p is due to Dor and Starbird [17]. The spaces

ℓ_p, $1 \leq p < 2$ are the only Banach spaces with a symmetric basis and a unique structure as a strictly 2-concave Banach lattice (and there is a dual statement for $p > 2$), as is clear from Proposition 3.a.5 of [33]. However, for r.i. function spaces on $[0,1]$ the situation is rather different. We show in Section 8 that under certain technical assumptions which apply to a wide range of spaces that the strictly 2-concave lattice structure on a separable r.i. space on $[0,1]$ is unique. The simplest such assumption (with the simplest proof) is that $p_X = 1$. However uniqueness also holds if $p_X \neq q_X$ or if X is p_X-convex. There are dual results for strictly 2-convex structures. It is to be noted that these results are of a somewhat different nature to the L_p−results which in the case $p > 2$ apply to any subspace and are proved by p-convexity and p-concavity considerations.

In Section 9 we pursue these considerations one step further by developing a simple criterion which allows us to conclude that a Banach lattice does not contain uniformly complemented ℓ_2^n's. This enables us to give examples of r.i. spaces on $[0,1]$ with unique structure as nonatomic Banach lattices, in addition to the known examples of L_1 and L_2. Roughly speaking an r.i. space which is very close to L_1 will have such a unique structure. For example the Orlicz space $L_F[0,1]$ has unique nonatomic structure when $F(t) \sim t(\log t)^\alpha$ for large t and $0 < \alpha < 1/2$. This example is already mentioned in [22] as an r.i. space on $[0,1]$ which is not isomorphic to an r.i. space on $[0,\infty)$.

Finally in Section 10 we consider embedding results in the spirit of [22] and give some small improvements to results from [22]. Let us draw attention to Theorem 10.9. It is shown in [22] that, if $1 < p < 2$ L_p is embeddable into any r.i. space on $[0,1]$ which contains $L(p,\infty)$ (i.e. weak L_p) by using p-stable random variables. Theorem 10.9 is motivated by an attempt to provide a converse; an r.i. space of Rademacher type p (e.g. the Lorentz space $L(p,q)$ where $p < q < \infty$) is a proper subspace of $L(p,\infty)$. In Theorem 10.7 we show that if L_p embeds into a Banach lattice of type p then it embeds as a sublattice.

We now give a brief description of Sections 2-6 where we develop our techniques to produce these results. Section 2 is mainly devoted to introducing terminology, concerning Köthe function spaces and Banach lattices. However we also discuss criteria for complementation of sublattices. In Section 3 we introduce the basic ideas on representing positive operators by generalized kernels (or random measures). These ideas have a long history having been explored in [24], [46], and [47], for example. In Section 4 we consider the basic situation when a Köthe function space X is known to be isomorphic to a complemented subspace of a Köthe function space Y. We associate to X and Y three positive operators P, Q, R defined on certain associated spaces. In Section 5 we concentrate on establishing, under suitable hypotheses, that these operators carry the information that X is complemented in Y (and, in particular, do not vanish). In Section 6 we study hypotheses which enable us to argue that the generalized kernels associated to these operators are given, a.e., by atomic measures. Thus by Section 7 we are ready to interpret the results.

Finally let us stress that Köthe function spaces and Banach lattices are (at least in the order-continuous case) essentially interchangeable notions, the former being simply a concrete version of the latter. It is frequently more convenient to deal with Köthe spaces but the reader should keep in mind that this involves no loss of generality.

2. Banach lattices and Köthe function spaces

Let us first recall that a Banach lattice X is said to be order-continuous if and only if every order-bounded increasing sequence is norm convergent (see [33] p.7). A Banach lattice which does not contain a copy of c_0 is automatically order-continuous but the converse is false. An atom in a Banach lattice is a positive element a so that $0 \leq x \leq a$ implies that $x = \alpha a$ for some $0 \leq \alpha \leq 1$. A Banach lattice is discrete if $\min(|x|, a) = 0$ for every atom a implies $x = 0$; it is nonatomic if it contains no atoms. The reader is referred to Aliprantis-Burkinshaw [3], Lindenstrauss-Tzafriri [33] or Schaefer [43] as a general reference for Banach lattices.

Let K be a Polish space (i.e. a separable complete metric space) and let μ be a σ-finite Borel measure on K; see Cohn [14] for background on measure theory. We refer to the pair (K, μ) as a Polish measure space; if μ is a probability measure then we say (K, μ) is a Polish probability space. We denote by $\mathcal{B}(K)$ the collection of Borel subsets of K; if E is a Borel set then χ_E denotes its indicator function. We denote by $L_0(\mu)$ the space of all Borel measurable functions on K, where we identify functions differing only on a set of measure zero; the natural topology of L_0 is convergence in measure on sets of finite measure. If $0 < p \leq 1$, an admissible p-norm is then a lower-semi-continuous map $f \to \|f\|$ from $L_0(\mu)$ to $[0, \infty]$ such that:

(a) $\|\alpha f\| = |\alpha| \|f\|$ whenever $\alpha \in \mathbf{R}$, $f \in L_0$.

(b) $\|f + g\|^p \leq \|f\|^p + \|g\|^p$, for $f, g \in L_0$.

(c) $\|f\| \leq \|g\|$, whenever $|f| \leq |g|$ a.e. (almost everywhere).

(d) $\|f\| < \infty$ for a dense set of $f \in L_0$,

(e) $\|f\| = 0$ if and only if $f = 0$ a.e.

If $p = 1$, we call $\| \ \|$ an admissible norm; an admissible quasinorm is an admissible p-norm for some $0 < p \leq 1$.

7

A quasi-Köthe function space on (K, μ) is defined to be a dense order-ideal X in $L_0(\mu)$ with an associated admissible quasinorm $\| \ \|_X$ such that if $X_{max} = \{f : \|f\|_X < \infty\}$ then either:

(1) $X = X_{max}$ (X is *maximal*) or:

(2) X is the closure of the simple functions in X_{max} (X is *minimal*).

We remark here that, for Köthe function spaces, X_{max} coincides with the maximal normed extension of X introduced in [1].

If $\| \ \|_X$ is a norm then X is called a Köthe function space. Notice that according to our description we consider $\| \ \|_X$ to be well-defined on L_0. Any order-continuous Köthe function space is minimal. Also any Köthe function space which does not contain a copy of c_0 is both maximal and minimal.

Given any Köthe function space X and $0 < p < \infty$ we define X_p (see [35]) to be the quasi-Köthe space of all f such that $|f|^p \in X$ with the associated admissible quasinorm $\|f\|_{X_p} = \||f|^p\|_X^{1/p}$. It is readily verified that $\| \ \|_{X_p}$ is an admissible p-norm when $0 < p < 1$ and an admissible norm when $p > 1$. We will primarily use the case $p = 1/2$, in this paper. We will also use the subscript $+$ to denote the positive cone in a variety of situations, e.g. $X_+ = \{f : f \in X, \ f \geq 0\}$.

If X is an order-continuous Köthe function space then X^* can be identified with the Köthe function space of all f such that:

$$\|f\|_{X^*} = \sup_{\|g\|_X \leq 1} \int |fg| \, d\mu < \infty.$$

X^* is always maximal.

If μ is a probability measure then we say following [22], that a Köthe function space X is *good* if $L_\infty \subset X \subset L_1$ and further for $f \in L_0$, $\|f\|_1 \leq \|f\|_X \leq 2\|f\|_\infty$. It is well-known that any separable order-continuous Banach lattice can be represented as (i.e. is isometrically lattice-isomorphic to) a good Köthe function space on some Polish probability space (K, μ) (see [22] and [36]).

In the case when X is nonatomic we can require that $K = [0, 1]$ and $\mu = \lambda$ is Lebesgue measure. Alternatively we can take $K = \Delta = \{-1, +1\}^{\mathbf{N}}$ to be the Cantor group and take μ to be normalized Haar measure on Δ which we again denote by λ.

To be more precise let μ is a nonatomic (or continuous) probability measure on the

Polish space K. Consider a Borel map $\sigma : K \to \Delta$ such that $\lambda \circ \sigma^{-1} = \mu$ and such that σ is essentially one-one (i.e. there is a Borel set $E \subset K$ so that σ is one-one on E and $\mu(E) = 1$); we shall say that σ is an essential isomorphism between (K, μ) and (Δ, λ). Under these circumstances it is easy to show that there is a Borel map $\tau : \Delta \to K$ so that $\sigma \circ \tau(t) = t$, λ–a.e. and $\tau \circ \sigma(s) = s$, μ–a.e. If X is a Köthe function space on K then we define $X \circ \sigma$ to be space of $f \in L_0(\lambda)$ so that $f \circ \sigma \in X$ with the associated norm $\|f\|_{X \circ \sigma} = \|f \circ \sigma\|_X$; clearly $X \circ \sigma$ is isometrically lattice-isomorphic to X.

LEMMA 2.1. *Let (K, μ) be a Polish probability space, where μ is nonatomic, and suppose (E_n) is any sequence of Borel subsets of K. Then there is an essential isomorphism $\sigma : (K, \mu) \to (\Delta, \lambda)$ such that for each n there is an open subset $W_n \subset \Delta$ so that $\sigma^{-1} W_n \subset E_n$ and $\lambda(W_n) = \mu(E_n)$.*

PROOF: We shall assume that the sequence (E_n) contains a base for topology of K. Let (F_n) be a sequence of Borel sets in which each E_n is repeated infinitely often. We will construct a sequence of Borel sets $G_n(+1)$ in K so that if $G_n(-1) = K \setminus G_n(+1)$ and $V(\epsilon_1, \ldots, \epsilon_n) = \cap_{k=1}^n G_k(\epsilon_k)$ then $\mu(V(\epsilon_1, \ldots, \epsilon_n)) = 2^{-n}$.

Assume $n \in \mathbf{N}$, and that $G_k(+1)$ has been chosen for $1 \le k \le n-1$. (In the case $n = 1$, let $V(\emptyset) = K$; the construction described below can then be suitably interpreted.) For each $\epsilon_1, \ldots, \epsilon_{n-1} = \pm 1$ consider $V(\epsilon_1, \ldots, \epsilon_{n-1}) \cap F_n$. Define a Borel subset of $V(\epsilon_1, \ldots, \epsilon_{n-1})$, $H(\epsilon_1, \ldots, \epsilon_{n-1})$ of measure 2^{-n} so that if $0 \le \mu(V(\epsilon_1, \ldots, \epsilon_{n-1}) \cap F_n) \le 2^{-n}$ then this set is contained in $H(\epsilon_1, \ldots, \epsilon_{n-1})$; in the other case when the measure exceeds 2^{-n} we require $H(\epsilon_1, \ldots, \epsilon_{n-1}) \subset V(\epsilon_1, \ldots, \epsilon_{n-1}) \cap F_n$, Finally let $G_n(+1) = \cup_{\epsilon_i = \pm 1} H(\epsilon_1, \ldots, \epsilon_{n-1})$. Then of course $V(\epsilon_1, \ldots, \epsilon_{n-1}, +1) = H(\epsilon_1, \ldots, \epsilon_{n-1})$.

Now for any Borel set $E \subset K$ let $Z_n(E)$ be the union of all sets $V(\epsilon_1, \ldots, \epsilon_n)$ which are not contained in either E or $K \setminus E$. Note that $Z_n(E)$ is monotone decreasing and that by construction $\mu(Z_n(F_n)) \le \frac{1}{2}\mu(Z_{n-1}(F_n))$. Hence we have $\lim_{N \to \infty} \mu(Z_N(E_n)) = 0$ for every $n \in \mathbf{N}$.

Now we define $\sigma : K \to \Delta$ by $\sigma(s) = (h_n(s))_{n=1}^\infty$ where $h_n = \chi_{G_n(+1)} - \chi_{G_n(-1)}$. It is clear that σ is Borel and that $\lambda \circ \sigma^{-1} = \mu$.

For each $\epsilon_1, \ldots, \epsilon_n = \pm 1$ let $\Delta(\epsilon_1, \ldots, \epsilon_n) = \{(d_k)_{k=1}^\infty : d_j = \epsilon_j \text{ for } 1 \le j \le n\}$ be the corresponding clopen subset of Δ. For each E_n define $U_{N,n}$ be the union of all

$\Delta(\epsilon_1, \ldots, \epsilon_N)$ such that $V(\epsilon_1, \ldots, \epsilon_N)$ is contained in E_n. Then $\sigma^{-1}U_{N,n} \subset E_n$. Further $\lambda(U_{N,n}) \geq \mu(E_n) - \mu(Z_N(E_n))$. Thus if we set $U_n = \cup_{N=1}^{\infty} U_{N,n}$ then U_n is open in Δ, $\sigma^{-1}U_n \subset E_n$ and $\lambda(U_n) = \mu(E_n)$.

We conclude by showing that σ is an essential isomorphism. Let $D = \cup_{n=1}^{\infty}(E_n \setminus \sigma^{-1}U_n)$. Then $\mu(D) = 0$. If $s_1, s_2 \in K \setminus D$ are distinct there exists n so that $s_1 \in E_n$ and $s_2 \notin E_n$. Then $\sigma s_1 \in U_n$; however if $\sigma s_2 \in U_n$ then $s_2 \in \sigma^{-1}U_n \subset E_n$. Thus σ is one-one on $K \setminus D$. ∎

Let us now restrict attention to quasi-Köthe function spaces on a compact metric space K with a probability measure μ. In this case we will say that X has the *density property* if whenever $f \in X_{max}$ with $\|f\|_X \leq 1$ then there is a sequence of continuous functions ϕ_n so that $\phi_n \to f$ in L_0 and $\|\phi_n\|_X \leq 1$. Notice that if X has the density property then so does X_p for any $0 < p < \infty$. It is also easy to see that any order-continuous Köthe function space containing L_∞ has the density property.

If X is a Köthe function space we shall say that X has the *strong density property* if both X and X^* have the density property. In some circumstances (e.g. if X is reflexive) the strong density property is automatic. However, to cover all cases, we will need to show that we can always assume it.

PROPOSITION 2.2. *If X is a good order-continuous Köthe function space on a Polish probability space (K, μ) where μ is nonatomic. Then there is an essential isomorphism $\sigma : K \to \Delta$ so that $X \circ \sigma$ has the strong density property. Thus any separable order-continuous nonatomic Banach lattice is isometrically lattice-isomorphic to a (minimal) good Köthe function space on (Δ, λ) with the strong density property.*

PROOF: Since X is order-continuous, it is necessarily separable. Using this it is easy to show that there is a sequence (g_n) of nonnegative simple functions on K so that $\|g_n\|_{X^*} \leq 1$ and for every $f \in X$ we have $\|f\|_X = \sup_n \int |fg_n|d\mu$. Clearly this last equation also extends to X_{max}.

Now, by Lemma 2.1, we can find an essential isomorphism $\sigma : (K, \mu) \to (\Delta, \lambda)$ so that there is a sequence h_n of simple lower-semi-continuous functions on Δ such that $h_n \circ \sigma = g_n$ a.e.

Clearly, as remarked above, $X \circ \sigma$ has the density property since it is order-continuous and contains L_∞. We also can find a sequence of nonnegative continuous functions ϕ_n with $\|\phi_n\|_{(X \circ \sigma)^*} \leq 1$ so that $\|f\|_X = \sup_n \int |f\phi_n| d\mu$ for $f \in X$.

Now suppose $h \geq 0$ and $\|h\|_{(X \circ \sigma)^*} \leq 1$. Fix any strictly positive v in $X \circ \sigma$. Assume h is not in the closure of the set Q of continuous functions in the ball of $(X \circ \sigma)^*$ for convergence in measure; then the same is true for the stronger topology induced by the weighted L_1-space $L_1(vd\lambda)$. By the Hahn-Banach theorem there exists w with $|w| \leq v$ a.e. such that

$$\int hw \, d\mu = \alpha > \beta = \sup_{\phi \in Q} \int \phi w \, d\mu.$$

Clearly

$$\sup_{\phi \in Q} \int \phi w \, d\mu = \sup_{\psi \in C(\Delta), |\psi| \leq 1} \sup_{\phi \in Q} \int \phi \psi w \, d\mu$$

so that

$$\beta = \sup_{\phi \in Q} \int |\phi||w| d\mu = \|w\|_X$$

and we have a contradiction.∎

Let X and Y be Banach lattices. A bounded linear operator $V : X \to Y$ is called a *lattice homomorphism* if $V(x_1 \vee x_2) = Vx_1 \vee Vx_2$ for every $x_1, x_2 \in X$. V is necessarily positive and it is automatic that the range of a lattice homomorphism is a sublattice of Y. If X is an order-continuous Köthe function space on (K_1, μ_1) and Y is a Köthe function space on (K_2, μ_2) it follows easily, using for example the ideas of [47] or Section 3 below, that every lattice homomorphism takes the form $Vf(s) = a(s)f(\sigma s)$ μ_2-a.e. where $a \in L_0(\mu_2)$ and $\sigma : K_2 \to K_1$ is a Borel map. (Of course the boundedness of V will impose additional constraints on a and σ.)

If X is a Köthe function space on (K_1, μ_1) and E is a Borel subset of K_1 of positive measure then we denote by X_E the band in X of f such that $f(s) = 0$ a.e. for $s \notin E$.

The following proposition is our main criterion for determining that a sublattice of a Köthe function space is complemented.

PROPOSITION 2.3. *Suppose X is an order-continuous Köthe function space on (K_1, μ_1) and Y is an order-continuous Köthe function space on (K_2, μ_2) where K_1, K_2 are Polish spaces and μ_1, μ_2 are σ-finite Borel measures. Suppose E is a Borel subset of K_1 of positive measure.*

(a) If X_E is lattice-isomorphic to a complemented sublattice of Y then there exist lattice-homomorphisms $U : X \to Y$ and $V : X^* \to Y^*$ of the form $Uf(s) = a(s)f(\sigma s)$ and $Vf(s) = b(s)f(\sigma s)$ where $a, b \in L_0(\mu_2)$ and $\sigma : K_2 \to K_1$ such that for some $\delta > 0$, whenever $F \subset E$ is a Borel set then

$$(*) \qquad\qquad \int_{\sigma^{-1}F} a(s)b(s) \, d\mu_2(s) \geq \delta\mu_1(F).$$

(b) Conversely suppose that there exist lattice homomorphisms $U : X \to Y_{max}$ and $V : X^* \to Y^*$ of the form $Uf = a(f \circ \sigma)$ and $Vf = b(f \circ \sigma)$ where for some $\delta > 0$ and every Borel set F contained in E we have $(*)$. Then X_E is lattice-isomorphic to a complemented sublattice of Y. More precisely there is a constant $C = C(\|U\|, \|V\|, \delta)$ so that X_E is C-lattice-isomorphic to a C-complemented sublattice of Y.

PROOF: (a) We can suppose the existence of a lattice-homomorphism $U : X \to Y$ and a bounded operator $T : Y \to X$ such that $TUf = f$ for $f \in X_E$. We assume $Uf = af \circ \sigma$ where $a \in L_0(\mu_2)$ is nonnegative and $\sigma : K_2 \to K_1$ is a Borel map. Consider the directed set $\mathcal{D}$ of all finite partitions $\mathcal{P} = \{G_1, \ldots, G_n\}$ of K_1 into Borel sets (directed by refinement). For any such $\mathcal{P}$ we define $R_{\mathcal{P}} : X^* \to Y^*$ by

$$R_{\mathcal{P}}f = \sum_{k=1}^{n} \chi_{\sigma^{-1}G_k} T^*(f\chi_{G_k}).$$

Then by a standard diagonalization argument (cf. Proposition 1.c.8, p.20 of [32]) we have $\|R_{\mathcal{P}}\| \leq \|T\|$. Hence by a compactness argument we can produce an operator $V_0 : X^* \to Y^*$, with $\|V_0\| \leq \|T\|$, so that for each $f \in X^*$, $V_0 f$ is a weak*-cluster point of the net $(R_{\mathcal{P}}(f))_{\mathcal{P} \in \mathcal{D}}$. It then follows easily that for any Borel set G and any $f \in X^*$ the function $V_0(f\chi_G)$ is supported in $\sigma^{-1}G$. Pick any strictly positive $w \in X^*$. Let $b_0 = (V_0 w)(w \circ \sigma)^{-1}$. Then it follows that $V_0(w\chi_G) = \chi_{\sigma^{-1}G} V_0 w = b_0((w\chi_G) \circ \sigma)$. It follows quickly that if $|f| \leq Mw$ for some constant M then $V_0 f = b_0 f \circ \sigma$. It is not difficult then to deduce that b_0 is independent of the original choice of w and that $V_0 f = b_0 f \circ \sigma$ for all $f \in X^*$.

Finally we let $b = |b_0|$ and $Vf = bf \circ \sigma$. It remains to verify $(*)$. If $F \subset E$ we use a theorem of Lozanovskii [36] to factorize $\chi_F = fg$ where $f \in X_{max,F}, g \in X_F^*$ with $f, g \geq 0$ and $\|f\|_X \|g\|_{X^*} = \mu_1(F) = \|\chi_F\|_1$. We then may select an increasing sequence F_m of Borel subsets of F so that $\cup F_m = F$ and $f\chi_{F_m} \in X$. For any partition $\mathcal{P} = \{G_1, \ldots, G_n\}$ we have

$$\int_{K_2} (U(f\chi_{F_m}))(R_\mathcal{P}g)d\mu_2 = \sum_{k=1}^{n} \int_{K_2} (U(f\chi_{F_m}))\chi_{\sigma^{-1}G_k} T^*(g\chi_{G_k})d\mu_2$$

$$= \sum_{k=1}^{n} \int_{K_2} (U(f\chi_{F_m \cap G_k}))T^*(g\chi_{G_k})d\mu_2$$

$$= \sum_{k=1}^{n} \int_{K_1} (TU(f\chi_{F_m \cap G_k}))g\chi_{G_k}d\mu_1$$

$$= \int_{K_1} f\chi_{F_m}g\, d\mu_1$$

$$= \mu_1(F_m).$$

Passing to the limit we obtain

$$\int_{K_2} (U(f\chi_{F_m}))V_0 g\, d\mu_2 = \mu_1(F_m)$$

which translates as

$$\int_{K_2} ab_0(f\circ\sigma)(g\circ\sigma)\chi_{\sigma^{-1}F_m}\, d\mu_2 = \mu_1(F_m).$$

Letting $m \to \infty$ we obtain (*) with $\delta = 1$.

(b) Notice that we do not require that U maps X into Y but rather into Y_{max}. To cope with this we suppose first that $v \in X$ is strictly positive and select $w_n \in Y$ so that $0 \le w_n \uparrow Uv$. Then let $a_n = w_n(Uv)^{-1}a$ and define $U_n f = a_n f \circ \sigma$. Since $U_n v \in Y$ it follows easily from order-continuity that $U_n(X) \subset Y$.

We next introduce Borel measures, ν_n, ν on K_1 by setting:

$$\nu_n(F) = \int_{\sigma^{-1}F} a_n b\, d\mu_2$$

$$\nu(F) = \int_{\sigma^{-1}F} ab\, d\mu_2.$$

We show that these measures are σ−finite and indeed μ_1−continuous. Indeed if $F \in \mathcal{B}(K_1)$ then utilizing Lozanovskii's theorem again we factor $\chi_F = fg$ where $f, g \ge 0$, $f \in X_{max}$, $g \in X_F^*$ and $\|f\|_X\|g\|_{X^*} = \mu_1(F)$. It is easy to see that U naturally extends to a lattice homomorphism from X_{max} into Y_{max} with the same norm. Hence

$$\nu(F) = \int_{\sigma^{-1}F} ab\, d\mu_2$$

$$= \int_{K_2} ab(f\circ\sigma)(g\circ\sigma)d\mu_2$$

$$\le \|U\|\|V\|\mu_1(F).$$

Now we can find by the Radon-Nikodym theorem $h_n, h \in L_0(\mu_1)_+$ so that

$$\nu(F) = \int_F h d\mu_1 \qquad \nu_n(F) = \int_F h_n d\mu_1 \qquad F \in \mathcal{B}(K_1).$$

Since $a_n \uparrow a$ a.e. it follows that $\nu_n(F) \uparrow \nu(F)$ for all F and hence $h_n \uparrow h$ μ_1−a.e.

Next we interpret (*). If $F \subset E$ we have $\nu(F) \geq \delta\mu_1(F)$ and hence $h \geq \delta\chi_E$ a.e. It now follows from Egoroff's theorem that we can partition E into Borel subsets (E_n) such that $h_n \geq \frac{1}{2}\delta\chi_{E_n}$. Finally we introduce another lattice homomorphism $W : X \to Y$ by setting

$$Wf = \sum_{n=1}^{\infty} U_n(f\chi_{E_n}) = (\sum_{n=1}^{\infty} a_n \chi_{\sigma^{-1}E_n})f \circ \sigma.$$

It is clear that W maps X into Y; in fact for each n, $U_n(f\chi_{E_n}) \in Y$ and the series $\sum f\chi_{E_n}$ converges to f by order-continuity. Further we have $\|W\| \leq \|U\|$.

We next consider the map $V^*W : X \to X^{**}$. We will show that this map may be considered as mapping into X_{max}. If fact if $f \in X_+$ and $g \in X_+^*$ then

$$\langle g, V^*Wf \rangle = \langle Wf, Vg \rangle$$
$$= \int_{K_2} (\sum_{n=1}^{\infty} a_n \chi_{\sigma^{-1}E_n})bf \circ \sigma g \circ \sigma \, d\mu_2$$
$$= \int_{K_1} \sum_{n=1}^{\infty} (h_n \chi_{E_n})fg \, d\mu_1$$

so that $V^*Wf = (\sum h_n \chi_{E_n})f$. Now by choice $\sum h_n \chi_{E_n} \geq \frac{1}{2}\delta\chi_E$ and so there is a lattice homomorphism $L : X_{max} \to X_{max}$ such that $LV^*Wf = \chi_E f$ for $f \in X$ and $\|L\| \leq 2/\delta$. It follows that $W(X_E)$ is a complemented sublattice of Y which is lattice-isomorphic to X_E.∎

If X is a Köthe function space on (K, μ) and if $1 \leq p < \infty$, we can define Köthe function spaces $X(\ell_p)$ and $X(L_p)$. The space $X(\ell_p)$ is a Köthe function space on $(K \times \mathbf{N}, \mu \times \pi)$ where π is counting measure on $\mathbf{N}$; an element F of $L_0(\mu \times \pi)$ can be identified with a sequence (f_n) of functions in $L_0(\mu)$ by $F(s, n) = f_n(s)$. $X(\ell_p)$ consists of all (f_n) such that $(\sum_{n=1}^{\infty} |f_n(s)|^p)^{1/p} \in X$ with the associated norm

$$\|F\|_{X(\ell_p)} = \|(\sum_{n=1}^{\infty} |f_n|^p)^{1/p}\|_X.$$

It is easily verified that $X(\ell_p)$ is minimal or maximal according as X is minimal or maximal, and is order-continuous when X is order-continuous. Similarly $X(L_p)$ is a Köthe function space on $(K \times [0,1], \mu \times \lambda)$ consisting of all $F = F(s,t)$ such that $(\int |F(s,t)|^p dt)^{1/p} \in X$ with the associated norm

$$\|F\|_{X(L_p)} = \|(\int |F(\cdot,t)|^p)^{1/p}\|_X.$$

It is now possible to define $X(\ell_p)$ and $X(L_p)$ for an arbitrary order-continuous Banach lattice by simply appealing to a concrete representation of X and this definition is unambiguous.

Our next result is a simple application of Proposition 2.3 which provides an interpretation of a result of Maurey [37] (cf. [33] Theorem 1.d.6 p. 49) concerning unconditional bases in Banach lattices. We state this result formally in order to provide motivation for our later results on nonatomic Banach lattices.

PROPOSITION 2.4. *Let X be a separable order-continuous atomic Banach lattice and suppose Y is an order-continuous Banach lattice. Suppose that X is isomorphic to a complemented subspace of Y. Then X is lattice-isomorphic to a complemented sublattice of $Y(\ell_2)$.*

PROOF: We suppose that X is a Köthe function space on $(\mathbf{N}, \pi)$ and that Y is a Köthe function space on (K, μ). Let $\chi_n = \chi_{\{n\}}$. Let $A : X \to Y$ and $B : Y \to X$ be linear operators such that $BA = I_X$ (the identity on X). Then X^* and Y^* can be identified as Köthe function spaces and $Y(\ell_2)^*$ can be identified with $Y^*(\ell_2)$.

Now by a well-known theorem of Krivine which we use heavily in the sequel ([33], p. 93 and [29]) we have that if $f \in X$ then for all $N \in \mathbf{N}$

$$\|(\sum_{n=1}^{N} |f(n)|^2 |A\chi_n|^2)^{1/2}\|_Y \leq K_G \|A\| \|f\|_X$$

(where K_G is the Grothendieck constant) and hence we can define a bounded lattice homomorphism $U : X \to Y_{max}(\ell_2) = Y(\ell_2)_{max}$ by $Uf(s,n) = f(n)A\chi_n(s)$. In a similar fashion since Y^* is already maximal we can define $V : X^* \to Y^*(\ell_2)$ by $Vf(s,n) = f(n)B^*\chi_n(s)$.

Thus $Uf = af \circ \sigma$ and $Vf = bf \circ \sigma$ where $\sigma : K \times \mathbf{N} \to \mathbf{N}$ is defined by $\sigma(s,n) = n$ and $a(s,n) = A\chi_n(s)$, $b(s,n) = B^*\chi_n(s)$. Now if Q is a finite subset of $\mathbf{N}$

$$\int_{\sigma^{-1}Q} a(s,n)b(s,n)d\mu = \sum_{n \in Q} \int (A\chi_n)(B^*\chi_n)d\mu = \pi(Q)$$

and so the proposition is an immediate consequence of Proposition 2.3.■

Let us now turn to rearrangement-invariant spaces (cf. [22],[33]). For any nonnegative $f \in L_0(K,\mu)$ we define its decreasing rearrangement $f^* \in L_0[0,\mu(K))$ by $f^*(t) = \sup\{x : \mu(f > x) < t\}$. Now let X be a Köthe function space on either $[0,\infty)$ or $[0,1]$ with Lebesgue measure. We say that X is a *rearrangement-invariant* (r.i.) space if $\|f\|_X = \|f^*\|_X$ for all $f \geq 0$, and if $\|\chi_{[0,1]}\|_X = 1$. If X is an r.i. space on $[0,\infty)$ (respectively, $[0,1]$) and (K,μ) is a Polish measure space (respectively, with $\mu(K) \leq 1$,) then we define $X(K,\mu)$ to be the set of $f \in L_0(\mu)$ such that $|f|^* \in X$ with $\|f\|_X = \||f|^*\|_X$. For example, it will be of some advantage to consider $X(\Delta,\lambda)$ in place of $X[0,1]$. We will only be interested in separable r.i. spaces which are necessarily minimal. It is then easy to see that $X(\Delta,\lambda)$ is a good Köthe function space with the strong density property (in fact, except in the exceptional case $X = L_1$, the closure of the simple functions in X^* is order-continuous which makes the conclusion immediate).

Consider a separable r.i. space X on $[0,1]$. In this case there is a certain rigidity in the structure of sublattices which can be expressed by the following obvious remark. Let (K,μ) be a nonatomic probability space. Suppose Σ is a sub-σ-algebra of $\mathcal{B}(K)$, and $E \in \Sigma$. Let $X_E(\Sigma)$ be the subspace of all $f \in X$ which are Σ-measurable and supported on E. Then $X_E(\Sigma)$ is a norm-one complemented sublattice of $X(K,\mu)$ by the conditional expectation operator. Further $X_E(\Sigma)$ is lattice-isomorphic to $X[0,1]$ if and only if μ is nonatomic when restricted to Σ. A slightly deeper observation is:

PROPOSITION 2.5. *Let X be an order-continuous Köthe function space on (K,μ) where K is a Polish space and μ is nonatomic Borel measure. Let Y be a separable r.i. space on $[0,1]$. Suppose X is isomorphic to a complemented sublattice of Y. Then there is a Borel subset E of K with $\mu(E) > 0$ so that X_E is lattice-isomorphic to Y.*

REMARKS: It is possible of course to restate this with X an order-continuous nonatomic Banach lattice. The conclusion would then be that there is a nontrivial band in X which is lattice isomorphic to Y. Also note that an exhaustion argument would allow us to partition K into countably many sets (E_n) such that X_{E_n} is lattice-isomorphic to Y. If X is also rearrangement-invariant and if $\mu(K) \leq 1$ then we obtain that X and Y are identical (up to equivalence of norm) as r.i. spaces. If we relax the assumption that Y is order-continuous the proposition is false. Indeed for $1 < p < \infty$, L_p is isomorphic to a complemented sublattice of $L(p, \infty)$ (weak L_p); this can be proved by considering the natural embedding of $L_p[0,1]$ into $L(p, \infty)([0,1]^2)$ defined by $Vf(s,t) = f(s)t^{-1/p}$. The argument of [27] Theorem 2.5 demonstrates complementation.

PROOF: We use Proposition 2.3. We may supppse the existence of lattice homomorphisms $U : X \to Y$ and $V : X^* \to Y^*$ of the forms $Uf = af \circ \sigma$ and $Vf = bf \circ \sigma$ where a, b are nonnegative functions in $L_0[0,1]$, $\sigma : [0,1] \to K$ is a Borel map and for every Borel subset F of K we have $\int_{\sigma^{-1}F} ab\, d\lambda \geq \mu(F)$.

Fix any $M < \infty$ so that the set $G = \{s \in [0,1] : M^{-1} \leq a(s), b(s) \leq M\}$ has positive measure. Then consider the measure ν on K defined by $\nu(F) = \lambda(\sigma^{-1}F \cap G)$. If $\mu(F) = 0$ then $U\chi_F = 0$ a.e. and hence $a\chi_{\sigma^{-1}F}=0$ a.e.; thus $\nu(F) = 0$. Further $\nu(K) = \lambda(G) > 0$. Thus, by an application of the Radon-Nikodym theorem there exists a constant $C > 0$ and a Borel set E in K so that $0 < \mu(E) < C^{-1}$ and whenever $F \subset E$ then $C^{-1}\mu(F) \leq \nu(F) \leq C\mu(F)$.

Now suppose $f \in X_E$ is nonnegative. Then $\|\chi_G f \circ \sigma\|_Y \leq M\|U\|\|f\|_X$ so that $\|f\|_Y \leq (C + 1)M\|U\|\|f\|_X$. Conversely pick $g \in X^*$ so that $\|g\|_{X^*} = 1$, $g \geq 0$ and $\int fg\, d\mu = \|f\|_X$. Then:

$$\|f\|_X \leq C \int fg\, d\nu$$
$$= C \int_G (f \circ \sigma)(g \circ \sigma)\, d\lambda$$
$$\leq CM \int_G Vg(f \circ \sigma)\, d\lambda$$
$$\leq CM\|V\|\|\chi_G f \circ \sigma\|_Y$$
$$\leq C(C + 1)M\|V\|\|f\|_Y.\ \blacksquare$$

On any r.i. space X on $[0,1]$ we define the dilation operators D_s for $0 < s < \infty$ by

$$D_s f(t) = f(t/s)$$

whenever $0 \leq t \leq \min(1, s)$ and $D_s f(t) = 0$ otherwise. The Boyd indices p_X and q_X are defined by

$$p_X = \lim_{s \to \infty} \frac{\log s}{\log \|D_s\|}$$

$$q_X = \lim_{s \to 0} \frac{\log s}{\log \|D_s\|}.$$

In general $1 \leq p_X \leq q_X \leq \infty$. If X is order-continuous, then X has an unconditional basis if and only if $1 < p_X \leq q_X < \infty$; in this case the Haar basis of X is an unconditional basis (see [33] p. 157-161). For convenience let us describe this basis for $X(\Delta, \lambda)$. Let $\Delta(\epsilon_1, \ldots, \epsilon_n) = \{d \in \Delta : d_j = \epsilon_j, \ 1 \leq j \leq n\}$. Let $\Delta(\emptyset) = \Delta$. For E a set of the form $\Delta(\epsilon_1, \ldots, \epsilon_n)$ where $n \geq 0$ we define the Haar function

$$h_E = \chi_{\Delta(\epsilon_1, \ldots, \epsilon_n, +1)} - \chi_{\Delta(\epsilon_1, \ldots, \epsilon_n, -1)}.$$

The functions (h_E) together with χ_Δ form the Haar system. If the Haar system is an unconditional basis then X has a representation as an atomic Banach lattice induced by the Haar system. We will call this the *Haar representation* of X and denote it by H_X.

A Köthe function space (or, more generally a quasi-Köthe function space) X is said to be $p-$convex (where $0 < p < \infty$) if there is a constant C such that for any $f_1, \ldots, f_n \in X$ we have

$$\|(\sum_{i=1}^{n} |f_i|^p)^{1/p}\|_X \leq C(\sum_{i=1}^{n} \|f_i\|_X^p)^{1/p}.$$

X is said to have an upper p-estimate if for some C and any disjoint $f_1, \ldots, f_n \in X$,

$$\|\sum_{i=1}^{n} f_i\|_X \leq C(\sum_{i=1}^{n} \|f_i\|_X^p)^{1/p}.$$

X is said to be $q-$concave $(0 < q < \infty)$ if for some $c > 0$ and any $f_1, \ldots, f_n \in X$ we have

$$\|(\sum_{i=1}^{n} |f_i|^q)^{1/q}\|_X \geq c(\sum_{i=1}^{n} \|f_i\|_X^q)^{1/q}.$$

X is said to have a lower q-estimate if for some $c > 0$ and any disjoint $f_1, \ldots, f_n \in X$,

$$\|\sum_{i=1}^{n} f_i\|_X \geq c(\sum_{i=1}^{n} \|f_i\|_X^q)^{1/q}.$$

Notice that a Köthe function space which satisfies a lower q-estimate is automatically both maximal and minimal since it cannot contain a copy of c_0.

A Banach lattice X is p-convex, satisfies an upper-estimate, is q-concave or satisfies a lower q-estimate according as any concrete representation of X as a Köthe function space has the same property.

A Banach space X is said to be of (Rademacher) type p $(1 \leq p \leq 2)$ if there is a constant C so that for any $x_1, \ldots, x_n \in X$,

$$\operatorname*{Ave}_{\epsilon_i = \pm 1} \| \sum_{i=1}^{n} \epsilon_i x_i \| \leq C (\sum_{i=1}^{n} \|x_i\|^p)^{1/p}$$

and X is of cotype q $(2 \leq q < \infty)$ if for some $c > 0$ and any $x_1, \ldots, x_n \in X$ we have

$$\operatorname*{Ave}_{\epsilon_i = \pm 1} \| \sum_{i=1}^{n} \epsilon_i x_i \| \geq c (\sum_{i=1}^{n} \|x_i\|^q)^{1/q}.$$

3. Positive operators

In this section we develop some machinery concerning positive operators and their representation by generalized kernels (or random measures). The basic idea of such representations has been explored in numerous articles. See for example [24] [26] [46] [47]. However we do develop some special results for future application.

Let K_1 be a compact metric space and let K_2 be a Polish space. By a *positive (generalized) kernel* or *random measure* we will mean a map $s \to \nu_s$ from K_2 into $\mathcal{M}_+(K_1)$ which is Borel for the weak*-topology on $\mathcal{M}(K_1)$ and the metric topology on K_2. It is easy to see that $s \to \nu_s$ is a positive generalized kernel then for every bounded Borel function f on K_1 the function $s \to \int f \, d\nu_s$ is a Borel map on K_2. Now suppose μ_2 is a σ−finite Borel measure on K_2. Given such a kernel we can induce a (continuous) positive operator $P : C(K_1) \to L_0(\mu_2)$ by the formula

$$Pf(s) = \int f(t) \, d\nu_s(t).$$

Conversely suppose $P : C(K_1) \to L_0(\mu_2)$ is any positive linear map. Select a linearly independent countable dense subset (f_n) in $C(K_1)$ and (selecting representatives) let $Pf_n = h_n$. Then if $a_1, \ldots, a_n$ are rational we have

$$|\sum_{i=1}^{n} a_i h_i(s)| \le P\chi_{K_1}(s) \max_{t \in K_1} |\sum_{i=1}^{n} a_i f_i(t)|$$

μ_2−a.e. It follows easily that for a suitable Borel subset E of K_2 with $\mu_2(K_2 \setminus E) = 0$ we can define Borel measures $\nu_s^P \in \mathcal{M}(K_1)$, for $s \in E$ with $\|\nu_s^P\| \le P\chi_{K_1}(s)$ and such that for every n $\int f_n d\nu_s^P = h_n(s)$. Off E we can set $\nu_s^P = 0$ for example. It follows then without difficulty that $\nu_s^P \ge 0$, μ_2- almost everywhere and that $s \to \nu_s^P$ is Borel; by redefining on a set of measure zero, ν_s^P is a positive kernel. Further ν_s^P is uniquely determined up to sets of μ_2−measure zero. Thus there is an identification of positive operators on $C(K_1)$ with positive kernels.

Now suppose that μ_1 is a probability measure on K_1. We will say that a positive operator P is *measure-continuous* if its kernel ν_s^P satisfies the condition that if E is a Borel subset of K_1 with $\mu_1(E) = 0$ then $\nu_s^P(E) = 0$ μ_2–a.e.

In general if P is measure continuous we extend the definition of Pf to all $f \in L_0(\mu_1)_+$ unambigously by the formula

$$Pf(s) = \int f(t) d\nu_s^P(t)$$

where Pf is a Borel function from K_2 into $[0, \infty]$. (It is immediate that if $f = g$ a.e. then $Pf = Pg$ a.e.)

LEMMA 3.1. *Let X be a quasi-Köthe function space on (K_1, μ_1) with the density property. Let Y be a maximal quasi-Köthe function space on K_2. Then if $P : C(K_1) \to L_0$ is a positive measure-continuous operator satisfying, for a suitable constant C, $\|Pf\|_Y \leq C\|f\|_X$ for all $f \in C(K_1)$ then P extends to a bounded positive operator $P : X \to Y$ defined by*

$$Pf(s) = \int f(t) \, d\nu_s^P(t) \quad \mu_2 - \text{a.e.}$$

and $\|P\|_{X \to Y} \leq C$.

If X is order-continuous and $L_\infty(\mu_1) \subset X$ any positive operator satisfying $\|Pf\|_Y \leq C\|f\|_X$ for all $f \in C(K_1)$ is measure-continuous.

PROOF: We define Pf as above for $f \in L_0(\mu_1)_+$. If $\|f\|_X \leq 1$ there is a sequence (g_n) with $g_n \in C_+(K_1)$ and $\|g_n\|_X \leq 1$ so that (g_n) converges μ_1–a.e. to f. Then (μ_2-a.e.) we have

$$Pf(s) \leq P(\liminf g_n)(s) = \int \liminf g_n \, d\nu_s^P \leq \liminf Pg_n(s)$$

so that

$$\|Pf\|_Y \leq \limsup_n \| \inf_{k \geq n} Pg_k\|_Y \leq C.$$

This estimate easily allows us to show that P is well-defined on X by the given formula with norm at most C.

For the final part if X is order-continuous, the density property is immediate. Further if E is closed in K_1 and $\mu_1(E) = 0$ then there exist $f_n \in C(K_1)$ with $\chi_E \leq f_n \leq 1$ with $f_n \to 0$ μ_1–a.e. Then $\|P\chi_E\|_Y \leq \|Pf_n\|_Y \to 0$. Hence $P\chi_E = 0$ a.e. and so P is measure-continuous.■

Next, if P and Q are two positive operators, $P, Q : C(K_1) \to L_0(\mu_2)$, we shall say that P is Q−continuous if ν_s^P is continuous with respect to ν_s^Q, μ_2−a.e. In general if P is not necessarily Q−continuous we may define its Q−continuous component $R = P_Q$ by the kernel

$$\nu_s^R = \sup_n (\nu_s^P \wedge n\nu_s^Q).$$

Here the lattice operations are in the space $\mathcal{M}(K_1)$. It is readily verified that ν_s^R is Borel and defines a positive kernel.

LEMMA 3.2. *If P is continuous with respect to Q there exists a Borel function g on $K_2 \times K_1$ such that for μ_2−a.e. $s \in K_2$,*

$$d\nu_s^P = g(s, t)d\nu_s^Q.$$

PROOF: We can define σ−finite Borel measures ν^P and ν^Q on $K_2 \times K_1$ by

$$\nu^P(E) = \int_{K_2} \int_{K_1} \chi_E(s, t) \, d\nu_s^P(t) \, d\mu_2$$

and

$$\nu^Q(E) = \int_{K_2} \int_{K_1} \chi_E(s, t) \, d\nu_s^Q(t) \, d\mu_2.$$

Then ν^P is ν^Q-continuous and hence there exists a Borel function g on $K_2 \times K_1$ such that $d\nu^P = g \, d\nu^Q$. Note that since g is Borel, for each s we also have that $t \to g(s, t)$ is Borel. Now if f is a continuous function on K_1, for every Borel subset F of K_2,

$$\int_F \int_{K_1} f(t) d\nu_s^P(t) \, d\mu_2 = \int_F \int_{K_1} f(t) g(s, t) \, d\nu_s^Q(t) \, d\mu_2.$$

Hence we obtain $\mu_2 - a.e.$

$$\int_{K_1} f(t) d\nu_s^P(t) = \int_{K_1} f(t) g(s, t) \, d\nu_s^Q(t).$$

Since $C(K_1)$ is separable, this leads to the lemma.∎

The following lemmas will be useful to us later.

LEMMA 3.3. *Let $P, Q, R, S : C(K_1) \to L_0(\mu_2)$ be positive operators. Suppose that P, Q, R are all S−continuous and that for every $f \in C_+(K_1)$ we have:*

$$P(f) \leq (Q(f))^{1/2}(R(f))^{1/2} \quad (\mu_2 - a.e.).$$

Then, if g_P, g_Q, g_R are Borel functions on $K_2 \times K_1$ such that $g_T(s, \cdot) . \nu_s^S$ is the kernel for T where $T = P, Q, R$, we have for μ−a.e. $s \in K$ that for ν_s^S−a.e. $t \in \Delta$,

$$g_P(s, t) \le (g_Q(s, t) g_R(s, t))^{1/2}.$$

Further we have that for any $f_1, f_2 \in C_+(K_1)$,

$$P(f_1 f_2) \le (Q(f_1^2))^{1/2}(R(f_2^2)^{1/2} \quad (\mu_2 - \text{a.e.}).$$

PROOF: We can exclude a set of μ_2− measure zero E so that for every $s \in K_2 \setminus E$, and for every continuous $f \in C_+(K_1)$ we have:

$$\int_{K_1} g_P(s, t) f(t) d\nu_s^S(t) \le \left(\int_{K_1} g_Q(s, t) f(t) d\nu_s^S(t) \right)^{1/2} \left(\int_{K_1} g_R(s, t) f(t) d\nu_s^S(t) \right)^{1/2}.$$

It follows that $s \in K \setminus E$ this equation holds for all bounded Borel functions f and hence the first part of the lemma follows. The last part follows directly from the Cauchy-Schwartz inequality. Assume $s \in K \setminus E$.

$$
\begin{aligned}
P(f_1 f_2)(s) &= \int g_P(s, t) f_1(t) f_2(t) \, d\nu_s^S(t) \\
&\le \left(\int g_Q(s, t)(f_1(t))^2 d\nu_s^S(t) \right)^{1/2} \left(\int g_R(s, t)(f_2(t))^2 d\nu_s^S(t) \right)^{1/2} \\
&= (Q(f_1^2)(s))^{1/2}(R(f_2^2)(s))^{1/2}. \blacksquare
\end{aligned}
$$

LEMMA 3.4. If P, Q, R satisfy

$$P(f) \le (Q(f))^{1/2}(R(f))^{1/2}$$

μ_2−a.e. whenever $f \in C(K_1)$ it also follows that

$$P(f) \le (Q_P(f))^{1/2}(R_P(f))^{1/2}$$

μ_2−a.e. for $f \in C(K_1)$.

PROOF: We may take $S = P + Q + R$ and apply Lemma 3.3. We obtain for for almost every s, $(g_P(s, t))^2 \le g_Q(s, t) g_R(s, t)$, ν_s^S−a.e. Let $E = \{(s, t) : g_P(s, t) > 0\}$. Let $g_Q' = \chi_E g_Q$

and $g'_R = \chi_E g_R$. It follows easily that $g'_Q \nu_s^S$ and $g'_R \nu_s^S$ are the kernels of Q_P and R_P respectively and that $(g_P(s,t))^2 \leq g'_Q(s,t) g'_R(s,t) \; \nu_s^S$–a.e. for μ_2–a.e. s.■

Finally we consider certain Köthe function spaces built using positive kernels. If X is a Köthe function space on some (K,μ) we recall that for $1 \leq p < \infty$ the Köthe function space $X(L_p)$ on $K \times [0,1]$ is defined to be the collection of all Borel functions f such that for μ–a.e. s, $f_s = f(s,\cdot) \in L_p[0,1]$ and $g \in X$ where $g(s) = \|f_s\|_p$, with $\|f\|_{X(L_p)} = \|g\|_X$.

Now suppose, as above K_1 is a compact metric space, K is a Polish space, μ is a σ–finite measure on K, and suppose $s \to \nu_s$, $(K \to \mathcal{M}_+(K_1))$ is any positive kernel, where $\psi(s) = \nu_s(K_1)$ is positive on a set of μ–positive measure. We define a σ–finite Borel measure μ_1 on $K \times K_1$ by

$$\mu_1(E) = \int_K \int_{K_1} \chi_E(s,t) d\nu_s(t) d\mu(s).$$

We define $X(L_p; \nu_s)$ to be the Köthe function space on $K_2 \times K_1$ of all f such that $f_s \in L_p(\nu_s)$ μ_2–a.e. and $g \in X$ where $g(s) = \|f_s\|_{L_p(\nu_s)}$. We define the norm by

$$\|f\|_{X(L_p; \nu_s)} = \|g\|_X.$$

LEMMA 3.5. $X(L_p; \nu_s)$ is isometrically lattice isomorphic to a 1-complemented sublattice of $X(L_p)$.

PROOF: We first remark that it suffices to consider the case when each ν_s is nonatomic. For we may embed $X(L_p; \nu_s)$ isometrically and naturally in $X(L_p; \rho_s)$ where $\rho_s = \nu_s \times \lambda \in \mathcal{M}_+(K_1 \times \Delta)$; then $X(L_p; \nu_s)$ is isometric to a 1-complemented sublattice of $X(L_p; \rho_s)$. Next, by using a Borel isomorphism we can assume that $K_1 = [0,1]$. Let us therefore restrict to this case.

Assume that $E = \{s : \psi(s) = \nu_s([0,1]) > 0\}$. We show that $X(L_p; \nu_s)$ is isomorphic to $X_E(L_p)$ (with obvious notation). Define a map $\phi : E \times [0,1] \to E \times [0,1]$ by $\phi(s,t) = \psi(s)^{-1}\nu_s(0,t)$. It is easy to show that the lattice isomorphism $f \to \psi^{1/p} f \circ \phi$ from $X_E(L_p; \nu_s)$ into $X_E(L_p)$ is isometric and surjective.■

4. The basic construction

Let, as before, $\Delta = \{-1,1\}^{\mathbf{N}}$ denote the Cantor group and let λ denote the standard Haar measure on Δ. For $n \in \mathbf{N}$ and $\epsilon_1, \ldots, \epsilon_n = \pm 1$ we let $\Delta(\epsilon_1, \ldots, \epsilon_n) = \{(d_k)_{k=1}^{\infty} : d_j = \epsilon_j \text{ for } 1 \leq j \leq n\}$. We let $\mathcal{B}(\Delta)$ denote the Borel sets of Δ and let $\mathcal{C}$ denote the algebra of clopen subsets of Δ. For each n, let $\mathcal{C}_n$ be the finite subalgebra of $\mathcal{C}$ generated by the atoms $\mathcal{A}_n = \{\Delta(\epsilon_1, \ldots, \epsilon_n) : \epsilon_i = \pm 1, \ 1 \leq i \leq n\}$. Let $CS(\Delta)$ denote the continuous simple functions on Δ (the linear span of functions of the form $\{\chi_E : E \in \mathcal{C}\}$.)

We now suppose that X is a good order-continuous Köthe function space on (Δ, λ) with the strong density property and that Y is a good order-continuous Köthe function space on (K, μ), where K is a Polish space and μ is a probability measure on K. We shall further suppose that there exist bounded linear operators $A : X \to Y$ and $B : Y \to X$ such that $BA = I_X$. For convenience we let $M = \max(\|A\|, \|B\|)$.

Before proceeding, we need a lemma which is already implicit in [22].

LEMMA 4.1. *Suppose that for each* $m \in \mathbf{N}$, $(\phi_{m,n})_{n=1}^{\infty}$ *is a sequence of non-negative functions in* $L_0(\mu)$ *so that* $\mathrm{co}\{\phi_{m,n} : n \in \mathbf{N}\}$ *is bounded. Then there is a sequence* $(\phi_m)_{m=1}^{\infty}$ *in* L_0 *and a strictly increasing sequence of natural numbers* $(n_k)_{k=1}^{\infty}$ *such that for each* $m \in \mathbf{N}$, *and for every subsequence* $f_{m,r}$ *of* (ϕ_{m,n_k}) $\lim_{r \to \infty} \frac{1}{r}(f_{m,1} + \cdots + f_{m,r})$ *converges a.e. to* ϕ_m.

PROOF: For each $m \in \mathbf{N}$ let K_m be an isomorphic copy of K and let $\tilde{K} = \cup_{m=1}^{\infty} K_m$ be the disjoint union. Let $\tilde{\mu}$ be the probability measure on $\tilde{K}$ given by $\tilde{\mu} = \sum_{m=1}^{\infty} \nu_m$ where ν_m is the measure $2^{-m}\mu$ transported to K_m. If $\psi_n(s) = \phi_{m,n}(s)$ for $s \in K_m$ it is easy to see that $\mathrm{co}\{\psi_n : n \in \mathbf{N}\}$ is bounded in $L_0(\tilde{\mu})$. By a theorem of Nikishin ([38],[39] or see [50], p.286 Example 6) there exists $w \in L_0(\tilde{\mu})$ with $w > 0$ a.e. and $\int w \, d\tilde{\mu} = 1$ such that $\mathrm{co}\{\psi_n : n \in \mathbf{N}\}$ is bounded in $L_1(w\tilde{\mu})$. By a well-known theorem of Komlos [28] we can find a $\psi \in L_1(w\tilde{\mu})$ and a strictly increasing sequence n_k so that every subsequence of (ψ_{n_k}) is Cesaro convergent a.e. to ψ and this gives the lemma.∎

Now if $E \in \mathcal{A}_n$ so that $E = \Delta(\epsilon_1, \ldots, \epsilon_n)$ we define the corresponding Haar function h_E by

$$h_E = \chi_{\Delta(\epsilon_1, \ldots, \epsilon_n, +1)} - \chi_{\Delta(\epsilon_1, \ldots, \epsilon_n, -1)}.$$

Let $CS_n(\Delta)$ denote the linear span of $\{\chi_E : E \in \mathcal{A}_n\}$.

For $n \in \mathbf{N}$ we define linear maps $P_n, Q_n, R_n : CS_n \to L_0(\mu)$ by setting $Q_n \chi_E = |Ah_E|^2$, $R_n \chi_E = |B^* h_E|^2$ and $P_n \chi_E = |Ah_E| . |B^* h_E|$ for $E \in \mathcal{A}_n$. As before we let K_G denote Grothendieck's constant.

LEMMA 4.2. *Suppose* $f, g \in CS_{n,+}$. *Suppose* $m \geq n$ *and that* $a_n, \ldots, a_m \geq 0$ *with* $a_n + \cdots + a_m = 1$. *Then:*

$$(1) \qquad \| \sum_{k=n}^{m} a_k Q_k f \|_{Y, 1/2} \leq K_G^2 M^2 \| f \|_{X, 1/2},$$

$$(2) \qquad \| \sum_{k=n}^{m} a_k R_k g \|_{Y^*, 1/2} \leq K_G^2 M^2 \| f \|_{X^*, 1/2},$$

$$(3) \qquad \sum_{k=n}^{m} a_k P_k (f^{1/2} g^{1/2}) \leq (\sum_{k=n}^{m} a_k Q_k f)^{1/2} (\sum_{k=n}^{m} a_k R_k g)^{1/2},$$

$$(4) \qquad \| \sum_{k=n}^{m} a_k P_k f \|_1 \leq K_G^2 M^2 \| f \|_X.$$

PROOF: (1) Let $f = \sum_{E \in \mathcal{A}_n} \alpha_E^2 \chi_E$, and let $a_k = b_k^2$. Then for $k \geq n$,

$$Q_k f = \sum_{E \in \mathcal{A}_n} \sum_{\substack{F \in \mathcal{A}_k \\ F \subseteq E}} \alpha_E^2 |Ah_F|^2.$$

Hence

$$\sum_{k=n}^{m} a_k Q_k f = \sum_{k=n}^{m} \sum_{E \in \mathcal{A}_n} \sum_{\substack{F \in \mathcal{A}_k \\ F \subseteq E}} b_k^2 \alpha_E^2 |Ah_F|^2.$$

Now by a theorem of Krivine (Theorem 1.f.14 of [33], [29])

$$\| \sum_{k=n}^{m} a_k Q_k f \|_{Y, 1/2} = \| \, | \sum_{k=n}^{m} a_k Q_k f |^{1/2} \|_Y^2$$

$$\leq K_G^2 M^2 \| \, | \sum_{k=m}^{n} \sum_{E \in \mathcal{A}_n} \sum_{\substack{F \in \mathcal{A}_k \\ F \subseteq E}} b_k^2 \alpha_E^2 \chi_F |^{1/2} \|_X^2$$

$$= K_G^2 M^2 \| f \|_{X, 1/2}.$$

The proof of (2) is similar. We turn to (3). Suppose $g = \sum_{E \in \mathcal{A}_n} \beta_E^2 \chi_E$. Then

$$\sum_{k=n}^{m} a_k P_k(f^{1/2} g^{1/2}) = \sum_{k=n}^{m} a_k \sum_{E \in \mathcal{A}_n} \sum_{\substack{F \in \mathcal{A}_k \\ F \subset E}} \alpha_E \beta_E |Ah_F| |B^* h_F|$$

$$\leq (\sum_{k=n}^{m} a_k Q_k f)^{1/2} (\sum_{k=n}^{m} a_k R_k g)^{1/2}.$$

Finally to prove (4) we take $g = \chi_\Delta$ and replace f by f^2. Then

$$\sum_{k=n}^{m} a_k P_k f \leq (\sum_{k=n}^{m} a_k Q_k(f^2))^{1/2} (\sum_{k=n}^{m} a_k R_k \chi_\Delta)^{1/2}.$$

Hence, since X is good and $\|\chi_\Delta\|_{X^*} \leq 1$,

$$\| \sum_{k=n}^{m} a_k P_k f \|_1 \leq \|(\sum_{k=n}^{m} a_k Q_k(f^2))^{1/2}\|_X \|(\sum_{k=n}^{m} a_k R_k \chi_\Delta)^{1/2}\|_{X^*}$$

$$= \| \sum_{k=n}^{m} a_k Q_k(f^2)\|_{X,1/2}^{1/2} \| \sum_{k=n}^{m} a_k R_k \chi_\Delta \|_{X^*,1/2}^{1/2}$$

$$\leq K_G^2 M^2 \|f\|_X. \blacksquare$$

Let us now define $P_n f = Q_n f = R_n f = 0$ whenever $f \in CS \setminus CS_n$. Then it follows easily from Lemma 4.1 and Lemma 4.2 that there is a strictly increasing sequence n_k so that for every $E \in \mathcal{C}$ we can define the $P\chi_E, \tilde{Q}\chi_E, \tilde{R}\chi_E$ to be those functions in L_0 which are the limits a.e. of the sequence of Cesaro means of any subsequence of, respectively, $(P_{n_k} \chi_E)_{k=1}^{\infty}$, $(Q_{n_k} \chi_E)_{k=1}^{\infty}$, or $(R_{n_k} \chi_E)_{k=1}^{\infty}$. It is clear from the mode of definition that $P, \tilde{Q}, \tilde{R}$ then extend to positive linear maps $P, \tilde{Q}, \tilde{R} : CS \to L_0(\mu)$. We further deduce, immediately,

LEMMA 4.3. For $f \in CS_+$,

(1) $$\|\tilde{Q}f\|_{Y,1/2} \leq K_G^2 M^2 \|f\|_{X,1/2}$$

(2) $$\|\tilde{R}f\|_{Y^*,1/2} \leq K_G^2 M^2 \|f\|_{X^*,1/2}$$

(3) $$\|Pf\|_1 \leq K_G^2 M^2 \|f\|_X.$$

It follows immediately that $P, \tilde{Q}$ and $\tilde{R}$ can extended to positive operators on $C(\Delta)$ obeying the same norm inequalities.

In equally simple manner we have (proving the result first for simple functions using Lemma 4.2(3)):

LEMMA 4.4. *If $f, g \in C_+(\Delta)$, then $P(fg) \leq (\tilde{Q}(f^2)\tilde{R}(g^2))^{1/2}$.*

Now it follows from Lemma 4.3(3) and Lemma 3.1 that P is measure-continuous and hence if we introduce its kernel ν_s^P we can extend P to a bounded operator $P : X \to L_1$ by the formula

$$Pf(s) = \int_\Delta f(t)d\nu_s^P(t).$$

We will define $Q = \tilde{Q}_P$ and $R = \tilde{R}_P$ so that both Q and R are P−continuous and hence also measure continuous. These then extend to operators $Q : X_{max,1/2} \to Y_{max,1/2}$ and $R : (X^*)_{1/2} \to (Y^*)_{1/2}$ (note here that both $X_{1/2}$ and $(X^*)_{1/2}$ have the density property.) By Lemma 3.2 there exist Borel functions k_Q and k_R on $K \times \Delta$ such that

$$Qf(s) = \int_\Delta k_Q(s,t)f(t)d\nu_s^P$$

and

$$Rf(s) = \int_\Delta k_R(s,t)f(t)d\nu_s^P.$$

We now state as a lemma:

LEMMA 4.5. *For μ−a.e. $s \in K$ we have*

$$k_Q(s,t)k_R(s,t) \geq 1$$

ν_s^P−a.e. and hence if $f \in X_+$, $g \in X_+^*$

$$P(fg) \leq (Q(f^2))^{1/2}(R(g^2))^{1/2}.$$

PROOF: It follows from Lemma 3.4 and Lemma 4.4 that for $f \in C_+(\Delta)$,

$$Pf \leq (Qf)^{1/2}(Rf)^{1/2}$$

and Lemma 3.3 gives the first part of the conclusion. The second part follows from the Cauchy-Schwartz inequality as in Lemma 3.3.∎

LEMMA 4.6. *P extends to a bounded operator $P : L_1(\lambda) \to L_1(\mu)$ with $\|P\| \leq K_G^2 M^2$.*

PROOF: Suppose $f \in L_{1,+}$ with $\|f\|_1 = 1$. Then, by a well-known factorization theorem due to Lozanovskii ([19],[36]) there exist $\phi \in X_{max,+}$ and $\psi \in X_+^*$ with $\|\phi\|_X = \|\psi\|_{X^*} = 1$

and $f = \phi\psi$. Now it follows that if we define $Pf(s) = \int f d\nu_s^P$ then $Pf \leq Q(\phi^2)^{1/2} R(\psi^2)^{1/2}$ almost everywhere and in particular $Pf(s) < \infty$ a.e. We further have

$$\|Pf\|_1 \leq \|Q(\phi^2)\|_{Y,1/2}^{1/2} \|R(\psi^2)\|_{Y^*,1/2}^{1/2}$$

so that

$$\|Pf\|_1 \leq K_G^2 M^2. \blacksquare$$

5. Lower estimates on P

This section is a continuation of Section 4 and we preserve the same notation. We now define a Borel measure π_P on Δ by setting

$$\pi_P(E) = \int \nu_s^P(E) d\mu(s) = \|P\chi_E\|_1.$$

It follows from Lemma 4.6 that $\pi_P \leq K_G^2 M^2 \lambda$.

LEMMA 5.1. *Suppose $\delta > 0$ and that F is a Borel subset of Δ with $\pi_P(F) < (1-\delta)\lambda(F)$. Then, given $\eta > 0$ and any finite collection $\{D_1, \ldots, D_m\}$ of clopen subsets of Δ, there exists a Borel subset F_0 of F with $\lambda(F_0) = \frac{1}{2}\lambda(F)$ and a Borel subset G of K with $\mu(G) < \eta$ so that if $\phi = 2\chi_{F_0} - \chi_F$ then $|\int_{D_j} \phi \, d\lambda| < \eta$ for $1 \leq j \leq m$ and*

$$\int_G (A\phi)(B^*\phi) d\mu > \delta\lambda(F).$$

PROOF: We first assume that F is clopen. Let $\tau = 4K_G^2 M^2 \eta^{-1}$. Then

$$\int \max(\tau - P\chi_F, 0) d\mu > \tau - (1-\delta)\lambda(F).$$

Recalling the definition of P, there exists a strictly increasing sequence of natural numbers n_k such that $F \in \mathcal{C}_{n_1}$ and (almost everywhere)

$$\max(\tau - P\chi_F, 0) \leq \liminf_{N \to \infty} \frac{1}{N} \sum_{k=1}^N \max(\tau - P_{n_k}\chi_F, 0).$$

Now we apply Fatou's Lemma to deduce that

$$\limsup_{k \to \infty} \int \max(\tau - P_{n_k}\chi_F, 0) d\mu > \tau - (1-\delta)\lambda(F)$$

so that

$$\liminf_{k \to \infty} \int \min(P_{n_k}\chi_F, \tau) d\mu < (1-\delta)\lambda(F).$$

We may now fix k so that $D_1, \ldots, D_m, F \in \mathcal{C}_{n_k}$ and

$$\int \min(P_{n_k} \chi_F, \tau) d\mu < (1 - \delta) \lambda(F).$$

Let $G = \{s : P_{n_k} \chi_F(s) > \tau\}$. Then $\tau \mu(G) \le \|P_{n_k} \chi_F\|_1 \le K_G^2 M^2 \|\chi_F\|_X \le 2K_G^2 M^2$ by Lemma 4.3. Hence $\mu(G) < \eta$. If $H = K \setminus G$ then

$$\int_H P_{n_k} \chi_F d\mu < (1 - \delta) \lambda(F).$$

Now let $\epsilon = \{\epsilon_E : E \in \mathcal{A}_{n_k}\}$ be a choice of signs. Let $g_\epsilon = \sum_{E \subset F} \epsilon_E h_E$. Then, averaging over all choices $\epsilon_E = \pm 1$,

$$\text{Ave} \, (Ag_\epsilon)(B^* g_\epsilon) = \sum_{E \subset F} (Ah_E)(B^* h_E) \le P_{n_k} \chi_F.$$

However,

$$\int_K \text{Ave} \, (Ag_\epsilon)(B^* g_\epsilon) d\mu = \lambda(F)$$

since $BA = I_X$. Hence we have

$$\int_G \text{Ave} \, (Ag_\epsilon)(B^* g_\epsilon) d\mu > \delta \lambda(F)$$

and there is a choice of signs ϵ_0 so that if $\phi = g_{\epsilon_0}$ then

$$\int_G (A\phi)(B^*\phi) d\mu > \delta \lambda(F).$$

By construction, it is immediate that $\int_{D_j} \phi \, d\lambda = 0$ for $1 \le j \le m$. If we define F_0 so that $\phi = 2\chi_{F_0} - \chi_F$ we have the conclusion of the lemma in the special case when F is clopen.

We now turn to the general case. If F is a Borel set of positive measure such that $\pi_P(F) < (1 - \delta)\lambda(F)$ then there exists $\gamma > 0$ so that if $\lambda(F \Delta F') < \gamma$ then $\pi_P(F') < (1 - \delta - \gamma)\lambda(F')$. We may then pick a sequence $F^{(n)} \in \mathcal{C}$ with $\lim \lambda(F^{(n)} \Delta F') = 0$ and $\lambda(F^{(n)} \Delta F') < \gamma$. By the first part we may then find clopen sets $F_0^{(n)}$ contained in $F^{(n)}$ with $\lambda(F_0^{(n)}) = \frac{1}{2}\lambda(F^{(n)})$ and Borel subsets G_n of K with $\mu(G_n) < \eta$ so that if $\phi_n = 2\chi_{F_0^{(n)}} - \chi_{F^{(n)}}$, then $\int_{D_j} \phi_n \, d\lambda = 0$ for $1 \le j \le m$, while

$$\int_{G_n} (A\phi_n)(B^*\phi_n) d\mu > (\delta + \gamma)\lambda(F^{(n)}).$$

We may further arrange, by taking $\mu(G_n)$ small enough that if $k < n$

$$\int_{G_n} |(A\phi_k)(B^*\phi_k)|d\mu < \frac{\gamma}{2^{n+1}}\lambda(F^{(k)}).$$

Thus if we set $G'_n = G_n \setminus \cup_{k>n} G_k$ then $\mu(G'_n) < \eta$ and

$$\int_{G'_n} (A\phi_n)(B^*\phi_n)d\mu > (\delta + \frac{1}{2}\gamma)\lambda(F^{(n)})$$

and the G'_n are disjoint.

Now, for each n there is a Borel subset $E_0^{(n)}$ of F with $\lambda(E_0^{(n)}) = \frac{1}{2}\lambda(F)$ and $\lambda(E_0^{(n)}\Delta F_0^{(n)}) \leq \lambda(F^{(n)}\Delta F)$. Let $\phi'_n = 2\chi_{E_0^{(n)}} - \chi_F$.

Notice that [3, p.202] the sequence $(\chi_{G'_n} A\phi_n)_{n=1}^{\infty}$ is weakly null. Also $\phi_n - \phi'_n$ converges to zero in measure and hence also in the Mackey topology $\tau(L_\infty, L_1)$ of uniform convergence on weakly compact subsets of L_1. Hence $\phi_n - \phi'_n$ also converges to zero for $\tau(X^*, X)$. Thus $B^*\phi_n - B^*\phi'_n$ converges to zero for $\tau(Y^*, Y)$. We deduce that

$$\lim_{n\to\infty} \int \chi_{G'_n} A\phi_n(B^*\phi'_n - B^*\phi_n)d\mu = 0.$$

We also have that $\lim \|\phi_n - \phi'_n\|_X = 0$ so that $\lim \|A\phi_n - A\phi'_n\|_Y = 0$. Hence

$$\lim_{n\to\infty} \int |A\phi'_n - A\phi_n||B^*\phi'_n|d\mu = 0.$$

Combining we obtain that for large enough n,

$$\int_{G'_n} (A\phi'_n)(B^*\phi'_n)d\mu > \delta\lambda(F).$$

It is also clear that $\lim \int_{D_j} \phi'_n d\lambda = 0$ for $1 \leq j \leq m$, and so the lemma follows, upon taking n large enough.∎

PROPOSITION 5.2. *Suppose Y contains no complemented sublattice lattice-isomorphic to ℓ_2. Then $\pi_P \geq \lambda$.*

REMARK: The hypothesis on Y is equivalent to the statement that no disjoint sequence spans a complemented Hilbertian subspace.

PROOF: Let us assume the existence of a Borel set F with $\pi_P(F) < (1 - \delta)\lambda(F)$ for some $\delta > 0$. Then, by induction, we may use Lemma 5.1 to construct a sequence of Borel subsets F_n of F with $\lambda(F_n) = \frac{1}{2}\lambda(F)$ and a sequence of Borel subsets G_n in K so that if $\phi_n = 2\chi_{F_n} - \chi_F$ then

$$\int_{G_n} (A\phi_n)(B^*\phi_n)d\mu > \delta\lambda(F)$$

while for $1 \leq k \leq n-1$,

$$\int_{G_n} |(A\phi_k)(B^*\phi_k)|d\mu < \frac{1}{2^{n+1}}\delta\lambda(F).$$

If we disjointify by setting $G'_n = G_n \setminus \cup_{k>n} G_k$ then

$$\int_{G'_n} (A\phi_n)(B^*\phi_n)d\mu > \frac{1}{2}\delta\lambda(F)$$

for all n.

Now let $w_n = \chi_{G'_n} A\phi_n$ and $w_n^* = \chi_{G'_n} B^*\phi_n$. then for any $a_1, \ldots, a_n$,

$$\begin{aligned}
\|\sum_{k=1}^n a_k|w_k|\|_Y &= \|(\sum_{k=1}^n |a_k|^2|w_k|^2)^{1/2}\|_Y \\
&\leq \|(\sum_{k=1}^n |a_k|^2|A\phi_k|^2)^{1/2}\|_Y \\
&\leq K_G M\|(\sum_{k=1}^n |a_k|^2|\phi_k|^2)^{1/2}\|_X \\
&= K_G M\|\chi_F\|_X (\sum_{k=1}^n |a_k|^2)^{1/2}.
\end{aligned}$$

In the above argument we use the result of Krivine ([33] Theorem 1.f.14). It now follows that there is a lattice homomorphism $L : \ell_2 \to Y$ with $Le_n = |w_n|$ where e_n is the canonical basis of ℓ_2.

Now if $f \in Y$ and $a_1, \ldots, a_n \in \mathbf{R}$,

$$\begin{aligned}
\left|\sum_{k=1}^n a_k \int |w_k^*|f\,d\mu\right| &\leq \|f\|_Y\|\sum_{k=1}^n a_k|w_k^*|\|_{Y^*} \\
&\leq K_G M\|\chi_F\|_{X^*}\|f\|_Y (\sum_{k=1}^n |a_k|^2)^{1/2}
\end{aligned}$$

by a similar calculation. It follows that there is a bounded positive operator $L' : Y \to \ell_2$ defined by

$$L'(f) = \left(\int |w_n^*|f\,d\mu\right)_{n=1}^{\infty}.$$

Now $L'Le_n = \alpha_n e_n$ where

$$\alpha_n = \int |w_n||w_n^*|d\mu \geq \frac{1}{2}\delta\lambda(F).$$

Thus $L'L$ is invertible and $[|w_n|]$ spans a complemented sublattice lattice-isomorphic to ℓ_2. ∎

We now turn to consideration of the case when X is rearrangement-invariant. We first reformulate Lemma 5.1.

LEMMA 5.3. *Let F be a Borel subset of Δ, with positive measure, such that $\pi_P(F) < (1-\delta)\lambda(F)$. Then there is a sequence $(F_n)_{n=1}^\infty$ of subsets of F with $\lambda(F_n) = \frac{1}{2}\lambda(F)$ and a sequence of Borel subsets $(G_n)_{n=1}^\infty$ of K with $\lim\mu(G_n) = 0$ so that*

$$\int_{G_n}(A\phi_n)(B^*\phi_n)d\mu > \delta\lambda(F),$$

where $\phi_n = 2\chi_{F_n} - \chi_F$, and $\lim\phi_n = 0$ weakly in X and weak* in X^*.

PROOF: In fact, it is immediate from Lemma 5.1 that one can choose (F_n) and (G_n) as above so that $\lim\int_D \phi_n\,d\lambda = 0$ for every clopen subset D of Δ. It follows from the order-continuity of X that (ϕ_n) will converge to zero weakly in X and weak* in X^*.

If X is an r.i. space then the Haar basis of X is the collection of functions $\chi_\Delta, \cup_n\{h_E : E \in \mathcal{A}_n\}$ taken in order. (The order of selection of h_E for fixed $\mathcal{A}_n$ is irrelevant).

LEMMA 5.4. *Let X be a separable r.i. space. Suppose the Haar basis of X satisfies the condition that there exists a constant C such that for every finitely non-zero set of scalars $\{a_E\}$ we have*

$$\|\sum_E a_E h_E\|_X \leq C\|(\sum_E |a_E|^2|h_E|^2)^{1/2}\|_X$$

$$\|\sum_E a_E h_E\|_{X^*} \leq C\|(\sum_E |a_E|^2|h_E|^2)^{1/2}\|_{X^*}.$$

Then the Haar basis of X is unconditional.

PROOF: It suffices ([33] p. 157) to show that the Boyd indices of X satisfy $p_X > 1$ and $q_X < \infty$. To do this it suffices to show that $q_X, q_{X^*} < \infty$. Consider then the dilation operator D_a for $0 < a < 1$, defined on $L_0[0,1]$ by $D_af(s) = f(s/a)$ for $0 \leq s \leq a$ and

$D_a f(s) = 0$ for $a < s \leq 1$. We require to show for suitable $a < 1$ we have $\|D_a\|_X$, $\|D_a\|_{X^*} < 1$ where the norms are computed with respect to $X[0,1]$ and $X^*[0,1]$.

To do this we pick $a = 2^{-n}$ where $n > C^2$. Suppose $f \in CS_+(\Delta)$ and let $f^* \in X[0,1]$ be the decreasing rearrangement of $|f|$. Suppose $f \in CS_m$. For each l let $r_l = \sum_{E \in \mathcal{A}_l} h_E$ (the Rademacher functions). Then

$$\|\sum_{l=m+1}^{m+n} fr_l\|_X \geq n\|D_a f^*\|_X.$$

However by the assumptions of the Lemma,

$$\|\sum_{l=m+1}^{m+n} fr_l\|_X \leq Cn^{1/2}\|f\|_X.$$

Combining these gives $\|D_a f^*\|_X \leq Cn^{-1/2}\|f\|_X$. This quickly gives that $\|D_a\|_X < 1$ and the estimate for $\|D_a\|_{X^*}$ is similar. ∎

PROPOSITION 5.5. *Suppose X is rearrangement-invariant and that either the Haar basis of X is not unconditional or that Y contains no complemented sublattice lattice-isomorphic to the Haar representation of X. Then $\pi_P \geq \lambda$.*

PROOF: Let us assume that $\pi_P(E) < \lambda(E)$ for some Borel set E. Then let $\beta = d\pi_P/d\lambda$. There exists $\delta > 0$ and a Borel set $F \subset \Delta$ of positive λ-measure so that $\beta(s) < 1 - \delta$ for $s \in F$. Thus if $F' \subset F$ is of positive measure then $\pi_P(F') < (1-\delta)\lambda(F')$.

Let $F_1 = F$. We shall determine a sequence of Borel sets $(F_n)_{n=1}^\infty$ in Δ and a sequence $(G_n)_{n=1}^\infty$ in K so that if we set $\phi_n = 2\chi_{F_n} - \chi_{F_{2n}}$ and let $\lambda_n = \lambda(F_n)$ we have:

(1): $F_n = F_{2n} \cup F_{2n+1}$ and $F_{2n} \cap F_{2n+1} = \emptyset$.

(2): $\lambda_{2n} = \lambda_{2n+1} = \frac{1}{2}\lambda_n$.

(3): For every $n \geq 1$:
$$\int_{G_n} (A\phi_n)(B^*\phi_n)\, d\mu > \delta\lambda_n.$$

(4): If $0 < k \leq n-1$ we have:
$$\left|\int_{G_n^k} (A\phi_k)(B^*\phi_n)\, d\mu\right| < \frac{\delta\lambda_n^2}{M2^{n+1}}$$

and

$$\left| \int_{G_n^k} (A\phi_n)(B^*\phi_k)\, d\mu \right| < \frac{\delta\lambda_n^2}{M2^{n+1}}$$

where $G_n^k = G_k \setminus \cup_{k<j<n} G_j$.

(5): If $1 \le k \le n-1$, and $E \subset G_n$ is any Borel set then

$$\|B(\chi_E A\phi_k)\|_1 < \frac{\delta\lambda_n^2}{M2^{n+2}}$$

and

$$\|A^*(\chi_E B^*\phi_k)\|_1 < \frac{\delta\lambda_n^2}{M2^{n+2}}.$$

Let us describe the inductive construction. Suppose $n \in \mathbf{N}$ and that we have determined $\{F_j : 1 \le j < 2n\}$, $\{\phi_j : 1 \le j < n\}$ and $\{G_j : 1 \le j < n\}$. Then, using Lemma 5.3, we can find a sequence of subsets $(F_{nl})_{l=1}^\infty$ of F_n with $\lambda(F_{nl}) = \frac{1}{2}\lambda_n$ and a sequence of Borel sets $(G_{nl})_{l=1}^\infty$ in K with $\lim_{l\to\infty} \mu(G_{nl}) = 0$ so that, if $\phi_{nl} = 2\chi_{F_{nl}} - \chi_{F_n}$ then $(\phi_{nl})_{l=1}^\infty$ is weakly null in X, weak*-null in X^* and

$$\int_{G_{nl}} (A\phi_{nl})(B^*\phi_{nl})d\mu > \delta\lambda_n.$$

Since Y is order-continuous,

$$\lim_{\mu(E)\to 0} \|\chi_E A\phi_j\|_Y = 0 \qquad 1 \le j \le n-1$$

and so

$$\lim_{\mu(E)\to 0} \|B(\chi_E A\phi_j)\|_X = 0 \qquad 1 \le j \le n-1.$$

Thus since the norm on X dominates the L_1-norm,

$$\lim_{\mu(E)\to 0} \|B(\chi_E A\phi_j)\|_1 = 0 \qquad 1 \le j \le n-1.$$

Although Y^* is not necessarily order-continuous, we do have

$$\lim_{\mu(E)\to 0} \chi_E B^*\phi_j = 0 \qquad 1 \le j \le n-1$$

for the Mackey topology $\tau(Y^*, Y)$. Since A^* is continuous for the respective Mackey topologies we also have

$$\lim_{\mu(E)\to 0} A^*(\chi_E B^*\phi_j) = 0 \qquad 1 \le j \le n-1$$

for the topology $\tau(X^*, X)$. It follows that

$$\lim_{\mu(E) \to 0} \|A^*(\chi_E B^* \phi_j)\|_1 = 0 \qquad 1 \leq j \leq n-1.$$

Utilizing these remarks, it is now possible to pick l large enough so that if we set $F_{2n} = F_{nl}$ and $F_{2n+1} = F_n \setminus F_{2n}$ and $G_n = G_{nl}$ then the inductive hypotheses are fulfilled.

Now let $H_n = G_n \setminus \cup_{k=n+1}^{\infty} G_k$. Then we can write

$$\int_{H_n} (A\phi_n)(B^*\phi_n)d\mu > \delta\lambda_n - \sum_{k=n+1}^{\infty} \int_{E_k} (A\phi_n)(B^*\phi_n)d\mu$$

where each E_k is a subset of G_k. Thus for each $k > n$

$$\left| \int_{E_k} (A\phi_n)(B^*\phi_n)d\mu \right| = \left| \int (B\chi_{E_k} A\phi_n)\phi_n d\lambda \right|$$
$$\leq \|B(\chi_{E_k} A\phi_n)\|_1$$
$$\leq \frac{\delta\lambda_k^2}{M2^{k+2}} < \frac{\delta\lambda_n}{2^{k+2}}.$$

Thus

$$(6) \qquad \int_{H_n} (A\phi_n)(B^*\phi_n)d\mu > \delta\lambda_n(1 - 2^{-(n+1)}) \geq \frac{\delta}{2}\lambda_n.$$

If $k > n$ then $H_k \subset G_k$ and by (5),

$$(7) \qquad \left| \int_{H_k} (A\phi_k)(B^*\phi_n)d\mu \right| \leq \|A^*(\chi_{H_k} B^*\phi_n)\|_1 \leq \frac{\delta\lambda_k^2}{M2^{k+2}}.$$

Similarly

$$(8) \qquad \left| \int_{H_k} (A\phi_n)(B^*\phi_k)d\mu \right| \leq \frac{\delta\lambda_k^2}{M2^{k+2}}.$$

If $k < n$ then

$$\int_{H_k} (A\phi_k)(B^*\phi_n)d\mu = \int_{G_n^k} (A\phi_k)(B^*\phi_n)d\mu - \sum_{j=n}^{\infty} \int_{E_j} (A\phi_k)(B^*\phi_n)d\mu$$

where, as before, $E_j \subset G_j$. Hence

$$\left| \int_{H_k} (A\phi_k)(B^*\phi_n)d\mu \right| \leq \frac{\delta\lambda_n^2}{M2^{n+1}} + \sum_{j=n}^{\infty} \|B(\chi_{E_j} A\phi_k)\|_1$$
$$\leq \frac{\delta\lambda_n^2}{M2^{n+1}} + \sum_{j=n}^{\infty} \frac{\delta\lambda_j^2}{M2^{j+2}}.$$

We conclude that

$$(9) \qquad |\int_{H_k}(A\phi_k)(B^*\phi_n)d\mu| \leq \frac{\delta\lambda_n^2}{M2^n}$$

and similarly

$$(10) \qquad |\int_{H_k}(A\phi_n)(B^*\phi_k)d\mu| \leq \frac{\delta\lambda_n^2}{M2^n}.$$

Now we define disjoint sequences (w_n) in Y and (w_n^*) in Y^* by $w_n = \chi_{H_n}A\phi_n$ and $w_n^* = \chi_{H_n}B^*\phi_n$. We complete the proof by showing that (ϕ_n) is unconditional, (w_n) is equivalent to (ϕ_n) and that $[w_n]$ is complemented in Y. From this it will follow immediately that Y contains a complemented sublattice lattice-isomorphic to the Haar representation of X.

There is a natural projection V of X onto $[\phi_n]$ given by

$$Vf = \sum_{n=1}^{\infty} \lambda_n^{-1}(\int_{\Delta} f\phi_n d\lambda)\phi_n.$$

Let

$$\alpha_n = \lambda_n^{-1}\int \phi_n(Bw_n)d\lambda$$

and observe that $\alpha_n \geq \delta/2$ for all n, by (6). Then

$$\int_{\Delta}\phi_n(Bw_k)d\lambda = \int_K (B^*\phi_n)\chi_{H_k}(A\phi_k)d\mu$$
$$= \int_{H_k}(A\phi_k)(B^*\phi_n)d\mu.$$

Thus we conclude that, using (7)-(10),

$$\|VBw_n - \alpha_n\phi_n\|_X \leq \sum_{k=1}^{n-1}\frac{\delta\lambda_n^2\lambda_k^{-1}\|\phi_k\|_X}{M2^n} + \sum_{k=n+1}^{\infty}\frac{\delta\lambda_k}{M2^k}\|\phi_k\|_X$$
$$\leq \sum_{k=1}^{n-1}\frac{\delta\lambda_n^2}{M2^n\lambda_k} + \frac{\delta\lambda_n}{M2^n}$$
$$\leq \frac{n\delta\lambda_n}{M2^n}$$
$$\leq \frac{n\delta\|\phi_n\|_X}{M2^n}.$$

Now $\|w_n\|_Y \leq \|A\phi_n\|_Y \leq M\|\phi_n\|_X$ and similarly, $\|w_n^*\|_{Y^*} \leq M\|\phi_n\|_{X^*}$. It follows that, by (6), $\delta\lambda_n/2 \leq M\|w_n\|_Y\|\phi_n\|_{X^*}$ and hence that $\delta\|\phi_n\|_X \leq 2M\|w_n\|_Y$. Combining with the estimates above now gives

$$\|VBw_n - \alpha_n\phi_n\|_X \leq \frac{n\|w_n\|_Y}{2^{n-1}}.$$

Notice that V^* is a projection of X^* onto the weak* closure of $[\phi_n]$; and very similar calculation gives that:

$$\|V^*A^*w_n^* - \alpha_n\phi_n\|_{X^*} \leq \frac{n\|w_n^*\|_{Y^*}}{2^{n-1}}.$$

Since $\sum k2^{-k} < \infty$ it follows easily that we can perturb VB to construct an operator $L : Y \to X$ such that $Lw_n = \alpha_n\phi_n$. Then we have, for any $a_1, \ldots, a_n$:

$$\|\sum_{k=1}^{n} a_k\phi_k\|_X = \|\sum_{k=1}^{n} a_k\alpha_k^{-1}Lw_k\|_X$$

$$\leq \|L\|\|\sum_{k=1}^{n} a_k\alpha_k^{-1}w_k\|_Y$$

$$\leq \|L\|\|(\sum_{k=1}^{n} |a_k|^2|\alpha_k|^{-2}|A\phi_k|^2)^{1/2}\|_Y$$

$$\leq K_G M\|L\|\|(\sum_{k=1}^{n} |a_k|^2|\alpha_k|^{-2}|\phi_k|^2)^{1/2}\|_X$$

$$\leq 2K_G M\|L\|\delta^{-1}\|(\sum_{k=1}^{n} |a_k|^2|\phi_k|^2)^{1/2}\|_X.$$

Precisely similar calculations show that there is a constant C so that for all $a_1, \ldots, a_n$

$$\|\sum_{k=1}^{n} a_k\phi_k\|_{X^*} \leq C\|(\sum_{k=1}^{n} |a_k|^2|\phi_k|^2)^{1/2}\|_{X^*}.$$

These estimates imply similar estimates for the Haar basis of X and hence by Lemma 5.4, the Haar basis of X is unconditional. Thus (ϕ_n) is also unconditional. Further the above calculations show the existence of a constant C so that for all $a_1, \ldots, a_n$

$$C^{-1}\|\sum_{k=1}^{n} a_k\phi_k\|_X \leq \|\sum_{k=1}^{n} a_k w_k\|_Y \leq C\|\sum_{k=1}^{n} a_k\phi_k\|_X$$

and

$$C^{-1}\|\sum_{k=1}^{n} a_k\phi_k\|_{X^*} \leq \|\sum_{k=1}^{n} a_k w_k^*\|_{Y^*} \leq C\|\sum_{k=1}^{n} a_k\phi_k\|_{X^*}.$$

Thus (w_n) is equivalent to (ϕ_n). Furthermore there is a bounded operator $U : X \to Y$ such that $U\phi_n = \alpha_n^{-1} w_n$ for every n. Clearly $U \circ L$ defines a projection of Y onto $[w_n]$. The sublattice generated by $(|w_n|)$ is then complemented and clearly Y contains a complemented sublattice lattice-isomorphic to the Haar representation of X.■

6. Reduction to the case of an atomic kernel

We first review the outcome of the methods of Sections 4 and 5.

THEOREM 6.1. *Suppose K is a Polish space and that μ is a probability measure on K. Let X be a good order-continuous function space on (Δ, λ) with the strong density property and let Y be a good order-continuous Köthe function space on (K, μ). Suppose $A : X \to Y$ and $B : Y \to X$ are bounded operators with $BA = I_X$ and let $M = \max(\|A\|, \|B\|)$.*

Assume further that either

(a) Y contains no complemented sublattice lattice-isomorphic to ℓ_2 or

(b) X is an r.i. space and (in the case when $1 < p_X \le q_X < \infty$) Y contains no complemented sublattice lattice-isomorphic to the Haar representation of X.

Then there exist positive (measure-continuous) operators $P, Q, R : C(\Delta) \to L_0(\mu)$ such that if $f \in L_0(\lambda)_+$

$$\|f\|_1 \le \|Pf\|_1 \le K_G^2 M^2 \|f\|_1$$

$$\|Qf\|_{Y,1/2} \le K_G^2 M^2 \|f\|_{X,1/2}$$

$$\|Rf\|_{Y^*,1/2} \le K_G^2 M^2 \|f\|_{X^*,1/2}$$

$$Pf \le (Qf)^{1/2}(Rf)^{1/2}$$

Furthermore Q, R are P-continuous.

PROOF: The operators P, Q, R were introduced in Section 4. The condition $\pi_P \ge \lambda$ is equivalent to $\|f\|_1 \le \|Pf\|_1$ for all $f \in L_0(\lambda)_+$ and so Propositions 5.2 and 5.5 are combined here.∎

In order to apply these results we need to consider situations when the kernel ν_s^P for P is automatically purely atomic for almost every s.

In general if $P : C(\Delta) \to L_0(\mu)$ is any measure-continuous positive operator with kernel ν_s^P (see [24], for example) we can find Borel maps $a_n : K \to [0, \infty)$ and $\sigma_n : K \to \Delta$, with $\sigma_m(s) \neq \sigma_n(s)$ whenever $m \neq n$, so that we have

$$\nu_s^P = \sum_{n=1}^{\infty} a_n(s)\delta_{\sigma_n(s)} + \rho_s$$

where ρ_s is a continuous (nonatomic) measure for every $s \in K$.

LEMMA 6.2. For $1 \leq j \leq N$ let $T_j : C(\Delta) \to L_0(\mu)$ be a measure-continuous positive operator. Let γ be any positive finite (μ-continuous) Borel measure on K. Let

$$\nu_s^{T_j} = \sum_{n=1}^{\infty} a_{n,j}(s)\delta_{\sigma_{n,j}(s)} + \rho_{j,s}$$

where each $a_{n,j}, \sigma_{n,j}$ is Borel, $\sigma_{n,j}(s) \neq \sigma_{m,j}(s)$, whenever $m \neq n$, and $\rho_{j,s}$ is continuous. Let $f_j \in L_0(\lambda)_+$ for $1 \leq j \leq N$. Define

$$u_j(s) = \left(\sum_{n=1}^{\infty} a_{n,j}^2(s)(f_j(\sigma_{n,j}(s)))^2 \right)^{1/2}$$

Then for any $D > N^{1/2}$ and any $\epsilon > 0$, there is a Borel subset F of K and a clopen subset E of Δ such that $\gamma(K \setminus F) < ND^{-2}\gamma(K)$ and for $1 \leq j \leq N$,

$$\chi_F|T_j(f_j\chi_E) - \frac{1}{2}T_j(f_j)| \leq Du_j + \epsilon T_j f_j.$$

PROOF: For each n let $\{G_{n,i} : 1 \leq i \leq 2^n\}$ be an enumeration of the atoms $\mathcal{A}_n$ of $\mathcal{C}_n$. For each n let

$$u_{n,j} = (\sum_{i=1}^{2^n}(T_j(f_j\chi_{G_{n,i}}))^2)^{1/2}.$$

Then clearly $u_{n,j}(s) \to u_j(s)$ for all $s \in K$. Now let r_k denote the Rademacher functions (which we regard as characters defined on the Cantor group Δ, i.e. $r_k(t) = t_k$) Then, by a simple estimate from Khintchine's inequality, for each $s \in K$ the set of $t \in \Delta$ such that

$$(*) \qquad |\sum_{i=1}^{2^n} r_i(t)T_j(f_j\chi_{G_{n,i}})(s)| \leq 2Du_{n,j}(s)$$

has λ-measure at least $1 - \frac{1}{4}D^{-2}$. It then follows from Fubini's theorem that we can find a Borel subset F_n' of K with $\gamma(F_n') > (1 - \frac{1}{2}ND^{-2})\gamma(K)$ and a fixed $t \in \Delta$ such that we have $(*)$ for every j and every $s \in F_n'$. Now let $E_n = \cup_{r_i(t)=1}G_{n,i}$. Then, for $1 \leq j \leq N$,

$$\chi_{F_n'}|T_j(f_j\chi_{E_n}) - \frac{1}{2}T_j f_j| \leq Du_{n,j}.$$

It remains to observe that as $u_{n,j} \to u_j$ pointwise and $u_{n,j} \leq T_j f_j$ we can apply Egoroff's theorem to find a sufficiently large n and a Borel set $F \subset F'_n$ with $\gamma(F) > (1-ND^{-2})\gamma(K)$ so that $Du_{n,j} \leq Du_j + \epsilon T_j f_j$ on F. The result then follows by taking $E = E_n$.

THEOREM 6.3. *Let* (K,μ) *be a Polish probability space. Suppose* X *is a good order-continuous function space on* (Δ, λ) *and let* Y *be an order-continuous function space on* (K,μ). *Suppose that* $M > 0$ *and that* $P, Q, R : C(\Delta) \to L_0(\mu)$ *are positive measure-continuous operators satisfying, for* $f \in L_0(\lambda)_+$,

$$Pf \leq (Qf)^{1/2}(Rf)^{1/2}$$

$$\|Qf\|_{Y,1/2} \leq M\|f\|_{X,1/2}$$

$$\|Rf\|_{Y^*,1/2} \leq M\|f\|_{X^*,1/2}.$$

Assume further that Q, R *are* P-*continuous. Then either:*

(a) L_2 *is lattice-isomorphic to a complemented sublattice of* X, *and, if further* X *is a separable r.i. space then* $X = L_2$.

or:

(b) There exists a sequence of Borel maps $\sigma_n : K \to \Delta$, *with* $\sigma_m(s) \neq \sigma_n(s)$ *for* $m \neq n$, *and there are sequences of nonnegative Borel functions* $a_n^P, a_n^Q, a_n^R \in L_0(\mu)$ *such that* $(\mu$-*a.e.*$)$

$$Pf(s) = \sum_{n=1}^{\infty} a_n^P(s)f(\sigma_n s)$$

$$Qf(s) = \sum_{n=1}^{\infty} a_n^Q(s)f(\sigma_n s)$$

$$Rf(s) = \sum_{n=1}^{\infty} a_n^R(s)f(\sigma_n s)$$

$$a_n^P(s)^2 \leq a_n^Q(s)a_n^R(s).$$

PROOF: We may write

$$\nu_s^P = \sum_{n=1}^{\infty} a_n^P(s)\delta_{\sigma_n s} + \rho_s$$

where ρ_s is almost everywhere a continuous measure. If $\rho_s = 0$ almost everywhere then we must have case (b) (using Lemma 3.3 for the last statement). Thus we shall assume that we do not have $\rho_s = 0$ almost everywhere and deduce case (a).

Let P_0 be the positive operator defined by

$$P_0 f(s) = \int f \, d\rho_s.$$

Then $P_0 f \leq (Qf)^{1/2}(Rf)^{1/2}$ for $f \in L_0(\lambda)_+$ and by an application of Lemma 3.4 we have $P_0 f \leq (Q_0 f)^{1/2}(R_0 f)^{1/2}$ where $Q_0 = Q_{P_0}$ and $R_0 = R_{P_0}$ are each P_0–continuous.

We shall define a Borel measure on K by $\gamma(F) = \int_F (P_0 \chi_\Delta) d\mu$. By assumption we have $\gamma(K) > 0$. Let $(\eta_n)_{n=0}^\infty$ be a sequence of positive reals such that $\sum_{n=0}^\infty 2^n \eta_n < \frac{1}{2} \min(1, \gamma(K))$. Let $G(\emptyset) = \Delta$. Then for each finite sequence $(\epsilon_1, \ldots, \epsilon_n)$ with $\epsilon_k = \pm 1$ we will define a clopen subset $G(\epsilon_1, \ldots, \epsilon_n)$ of Δ and a Borel subset $F(\epsilon_1, \ldots, \epsilon_n)$ of K so that

$$\chi_{G(\epsilon_1,\ldots,\epsilon_n)} = \chi_{G(\epsilon_1,\ldots,\epsilon_n,1)} + \chi_{G(\epsilon_1,\ldots,\epsilon_n,-1)}$$

$$\gamma(K \setminus F(\epsilon_1, \ldots, \epsilon_n)) < \eta_n$$

$$\chi_{F(\epsilon_1,\ldots,\epsilon_n)} |T(\chi_{G(\epsilon_1,\ldots,\epsilon_n,1)} - \tfrac{1}{2} T \chi_{G(\epsilon_1,\ldots,\epsilon_n)})| \leq \eta_n T \chi_{G(\epsilon_1,\ldots,\epsilon_n)}$$

where T is P_0, Q_0 or R_0. It is clear that this can be done by repeated application of Lemma 6.2 for suitably large D (in this case $u_j = 0$).

We now let $F = \cap_{n=1}^\infty \cap_{\epsilon_k = \pm 1} F(\epsilon_1, \ldots, \epsilon_n)$, so that $\gamma(K \setminus F) \leq \sum 2^n \eta_n < \frac{1}{2} \gamma(K)$. Thus $\gamma(F) > \frac{1}{2} \gamma(K) > 0$. Further if $T = P_0, Q_0$ or R_0

$$\chi_F |T \chi_{G(\epsilon_1,\ldots,\epsilon_n)} - \frac{1}{2^n} T \chi_\Delta| \leq \left(\sum_{k=0}^{n-1} 2^k \eta_k \right) 2^{-n} T \chi_\Delta$$

$$\leq 2^{-(n+1)} T \chi_\Delta$$

so that

$$(*) \qquad \chi_F T(\chi_{G(\epsilon_1,\ldots,\epsilon_n)}) \leq \frac{3}{2} 2^{-n} \chi_F T \chi_\Delta.$$

Consider the measure defined on Δ by $\pi_F(H) = \int_F P_0 \chi_H \, d\mu$. Then π_F is nonzero and $0 \leq \pi_F \leq M\lambda$, so that there is a Borel set E of positive measure in Δ so that for some $c > 0$, we have $\pi_F(E_0) \geq c\lambda(E_0)$ for every Borel set $E_0 \subset E$.

Now we define a continuous map $\sigma : \Delta \to \Delta$ so that $\sigma(s) = (\epsilon_k)_{k=1}^\infty$ whenever $s \in \cap_{n=1}^\infty G(\epsilon_1, \ldots, \epsilon_n)$. We then define the lattice homomorphism $U : C(\Delta) \to L_0(\lambda)$ by $Uf = \chi_E(f \circ \sigma)$.

At this stage we observe that (*) implies that if $T = P_0, Q_0$ or R_0 then, for any $f \in C(\Delta)_+$ we have

$$(**) \qquad\qquad \chi_F T(f \circ \sigma) \le \frac{3}{2}\left(\int_\Delta f\, d\lambda\right)\chi_F T\chi_\Delta.$$

For $f \in C(\Delta)_+$ we have $\chi_E(f \circ \sigma) \in X$ and so there exists $g \in X^*$ with $\|g\|_{X^*} = 1$ and $g \ge 0$ so that

$$\int_E g(f \circ \sigma)d\mu_1 = \|\chi_E(f \circ \sigma)\|_X.$$

Now

$$P_0(g(f \circ \sigma)) \le (Q_0(f \circ \sigma)^2)^{1/2}(R_0 g^2)^{1/2},$$

and

$$\begin{aligned}
c\|\chi_E(f \circ \sigma)\|_X = c\int_E g(f \circ \sigma)d\lambda \\
\le \int_E g(f \circ \sigma)d\pi_F \\
= \int_F P_0(g(f \circ \sigma))d\mu.
\end{aligned}$$

Further

$$\|(R_0 g^2)^{1/2}\|_{Y^*} = \|R_0 g^2\|_{Y^*,1/2}^{1/2} \le M^{1/2}\|g^2\|_{X^*,1/2}^{1/2} \le M^{1/2}.$$

Hence

$$\begin{aligned}
\|Uf\|_X = \|\chi_E(f \circ \sigma)\|_X \\
\le \frac{M^{1/2}}{c}\|(\chi_F Q_0(f \circ \sigma)^2)^{1/2}\|_Y \\
= \frac{M^{1/2}}{c}\|\chi_F Q_0(f \circ \sigma)^2\|_{Y,1/2}^{1/2} \\
\le \left(\frac{3M}{2c^2}\right)^{1/2}\|\chi_F Q_0\chi_\Delta\|_{Y,1/2}^{1/2}\|f\|_2
\end{aligned}$$

where we use the fact that

$$\chi_F Q_0(f \circ \sigma)^2 \le \frac{3}{2}\left(\int_\Delta f^2\, d\lambda\right)\chi_F Q_0\chi_\Delta$$

which follows from (**).

It follows that U extends to a bounded lattice homomorphism from $L_2(\Delta, \lambda)$ into X, and this extension is also given by $f \rightarrow \chi_{Ef} \circ \sigma$. A very similar calculation shows that U is a bounded lattice homomorphism of L_2 into X^*. We now seek to apply Proposition 2.3. In order to this we must consider the measure on Δ defined by $\nu(H) = \lambda(\sigma^{-1}H \cap E)$. It follows from Proposition 2.3 that if we can find a Borel subset H_0 of Δ of positive $\lambda-$measure and $\delta > 0$ so that if $H \subset H_0$ then $\nu(H) \geq \delta\lambda(H)$ then X contains a complemented sublattice lattice-isomorphic to L_2. To see this we need only show that ν is $\lambda-$continuous and nonzero and then use the Radon-Nikodym Theorem. First $\nu(\Delta) = \lambda(E) > 0$. Second if $\lambda(H) = 0$ then $\|U\chi_H\|_X = 0$ and hence $\nu(H) = 0$. The proof is then complete.

Finally if X is a separable r.i. space and contains L_2 as a complemented sublattice then $X = L_2$ by Proposition 2.5.∎

THEOREM 6.4. *Under the hypotheses of Theorem 6.3 assume that case (b) holds and that for some $\alpha > 0$ we have*

$$\int Pf \, d\mu \geq \alpha \int f \, d\lambda$$

for all $f \in L_1(\lambda)_+$. Then either:

(c) There is a constant C so that for each $n \in \mathbf{N}$, ℓ_2^n is C-lattice isomorphic to a C-complemented sublattice of X,

or

(d) There exists a constant $c > 0$ so that for $f \in L_1(\lambda)_+$,

$$\int_K \sup_n |a_n^P f \circ \sigma_n| \, d\mu \geq c \int_\Delta f \, d\lambda.$$

PROOF: We proceed by assuming that (d) fails. We may further assume without loss of generality that $(a_n^P)^2 = a_n^Q a_n^R$ almost everywhere (for otherwise we may simply redefine Q).

Let us assume that $N \in \mathbf{N}$ is a power of two, say $N = 2^l$. Let $[N]$ be the set $\{1, 2, \ldots, N\}$ regarded as equipped with normalized counting measure. We will construct lattice homomorphisms $U : L_2([N]) \rightarrow X_{max}$ and $V : L_2([N]) \rightarrow X^*$.

To start we observe from the negation of (d) that there exists $f \geq 0$ in $L_1(\lambda)$ so that $\|f\|_1 = 1$ but

$$\int_K \sup_n |a_n^P f \circ \sigma_n| \, d\mu < N^{-4}.$$

We factorize $f = gh$ where $g \in X_{max,+}$, $h \in X_+^*$, $\|g\|_X = \|h\|_{X^*} = 1$. Let $\beta = 16M\alpha^{-2}N^{-4}$. Then for $s \in K$, let $\mathcal{N}_R = \mathcal{N}_R(s)$ be the set of $n \in \mathbf{N}$ for which we have

$$\beta a_n^R(s)(h(\sigma_n(s))^2 Q(g^2)(s) < (a_n^P(s))^2 (f(\sigma_n(s))^2.$$

We observe that, for each n, the set $\{s : n \in \mathcal{N}_R(s)\}$ is a Borel set.

It follows from the Cauchy-Schwarz inequality that

$$\sum_{n \in \mathcal{N}_R} a_n^P(s) f(\sigma_n(s)) \le (\sum_{n \in \mathcal{N}_R} a_n^Q(s)(g(\sigma_n(s))^2)^{1/2} (\sum_{n \in \mathcal{N}_R} a_n^R(s)(h(\sigma_n s)^2)^{1/2}$$

$$\le \beta^{-1/2} (\sum_{n \in \mathcal{N}_R} (a_n^P(s)f(\sigma_n(s))^2)^{1/2}$$

$$\le \beta^{-1/2} (Pf(s))^{1/2} (\sup_n |a_n^P(s)f(\sigma_n(s))|)^{1/2}.$$

Thus

$$\int \sum_{n \in \mathcal{N}_R} a_n^P f \circ \sigma_n \, d\mu \le \beta^{-1/2} \|Pf\|_1^{1/2} \|\sup_n a_n^P f \circ \sigma_n\|_1^{1/2} \le \beta^{-1/2} M^{1/2} N^{-2} \le \frac{\alpha}{4}.$$

We can repeat the same construction for Q; precisely, we let $\mathcal{N}_Q(s)$ be the set of n such that

$$\beta a_n^Q(s) g(\sigma_n s)^2 R(g^2)(s) < (a_n^P(s)f(\sigma_n s))^2.$$

Then

$$\int \sum_{n \in \mathcal{N}_Q} a_n^P f \circ \sigma_n d\mu \le \frac{\alpha}{4}.$$

Thus if we set $\mathcal{M}(s) = \mathbf{N} \setminus (\mathcal{N}_Q(s) \cup \mathcal{N}_R(s))$ then

$$\int \sum_{n \in \mathcal{M}} a_n^P f \circ \sigma_n d\mu \ge \frac{\alpha}{2}.$$

We also have

$$\sup_{n \in \mathcal{M}} a_n^Q (g \circ \sigma_n)^2 \le \beta Q(g^2)$$

$$\sup_{n \in \mathcal{M}} a_n^R (h \circ \sigma_n)^2 \le \beta R(g^2).$$

We now modify our operators P, Q, R. We define P', Q', R' by

$$P'\phi = \sum_{n \in \mathcal{M}} a_n^P \phi \circ \sigma_n$$

$$Q'\phi = \sum_{n \in \mathcal{M}} a_n^Q \phi \circ \sigma_n$$

$$R'\phi = \sum_{n \in \mathcal{M}} a_n^R \phi \circ \sigma_n.$$

Notice that we still have that for $\phi \geq 0$, $P'\phi \leq (Q'\phi)^{1/2}(R'\phi)^{1/2}$.

Let

$$u = \Big(\sum_{n \in \mathcal{M}} (a_n^Q (g \circ \sigma_n)^2)^2 \Big)^{1/2}$$

$$v = \Big(\sum_{n \in \mathcal{M}} (a_n^R (h \circ \sigma_n)^2)^2 \Big)^{1/2}.$$

Then

$$u \leq \big(\sup_{n \in \mathcal{M}} a_n^Q (g \circ \sigma_n)^2 \big)^{1/2} Q'(g^2)^{1/2} \leq \beta^{1/2} Q(g^2),$$

and this yields the estimate

$$\|u\|_{Y,1/2} \leq \beta^{1/2} M \leq 4M^{3/2} \alpha^{-1} N^{-2}.$$

In an exactly similar fashion we obtain

$$\|v\|_{Y^*,1/2} \leq 4M^{3/2} \alpha^{-1} N^{-2}.$$

We next define a Borel measure γ on K by

$$\gamma(E) = \int_E P'f \, d\mu,$$

so that $\gamma(K) \geq \alpha/2$.

By iterating Lemma 6.2 with $D = 4N^{1/2}$ and $\epsilon = N^{-1}/2$ we may find a Borel subset F of K with $\gamma(K \setminus F) < 3D^{-2}N\gamma(K) < \gamma(K)/2$, and a Borel partition $\{E_1, \ldots, E_N\}$ of Δ such that

$$\chi_F |Q'(g^2 \chi_{E_j}) - N^{-1} Q'(g^2)| \leq 2Du + 2\epsilon Q'(g^2)$$

$$\chi_F |R'(h^2 \chi_{E_j}) - N^{-1} R'(h^2)| \leq 2Dv + 2\epsilon R'(h^2)$$

for $j = 1, 2, \ldots, N$. Thus

$$\chi_F Q'(g^2 \chi_{E_j}) \leq \chi_F(\frac{2}{N} Q'(g^2) + 2Du)$$

$$\chi_F R'(h^2 \chi_{E_j}) \leq \chi_F(\frac{2}{N} Q'(h^2) + 2Dv).$$

Now consider the measure π defined on Δ by $\pi(E) = \int_F P'(\chi_E) d\mu$. Then $\pi \leq M\lambda$, and so there is a Borel function θ with $0 \leq \theta \leq M$ and $\pi(E) = \int_E \theta d\lambda$ for all Borel sets E in Δ. If $\phi \in L_1(\lambda)$ then

$$\int_F P'\phi \, d\mu = \int \theta\phi \, d\lambda.$$

In particular

$$\int \theta f \, d\lambda = \int_F P' f \, d\mu = \gamma(F) \geq \frac{\alpha}{4}.$$

Hence if $G = \{s : \theta(s) \geq \alpha/8\}$, then

$$\int_G \theta f \, d\lambda \geq \frac{\alpha}{8}.$$

We now can construct $U : L_2([N]) \to X$ and $V : L_2([N]) \to X^*$ as promised. We let $\tau : \Delta \to [N]$ be defined by $\tau(s) = j$ where $s \in E_j$. Let

$$U\phi(s) = \phi(\tau s) g(s) \chi_G(s)$$

$$V\phi(s) = \phi(\tau s) h(s) \chi_G(s).$$

If $\phi \in L_2[N]_+$, there exists $\psi \in X_+^*$ supported on G with $\|\psi\|_{X^*} = 1$ and such that $\int \psi U\phi \, d\lambda = \|U\phi\|_X$. Thus:

$$\|U\phi\|_X = \int_G \psi U\phi \, d\lambda$$

$$\leq \frac{8}{\alpha} \int_G \theta\psi U\phi \, d\lambda$$

$$= \frac{8}{\alpha} \int_F P'(\psi U\phi) \, d\mu$$

$$\leq \frac{8}{\alpha} \int_F (Q'((U\phi)^2))^{1/2} (R'(\psi^2))^{1/2} d\mu$$

$$\leq \frac{8}{\alpha} \|\chi_F Q'((U\phi)^2)\|_{Y,1/2}^{1/2} \|R'(\psi^2)\|_{Y^*,1/2}^{1/2}$$

$$\leq \frac{8M^{1/2}}{\alpha} \|\chi_F Q'((U\phi)^2)\|_{Y,1/2}^{1/2}.$$

However we also have

$$\chi_F Q'((U\phi)^2) \le (\frac{1}{N}\sum_{k=1}^{N}\phi(k)^2)\chi_F(2Q'(g^2) + 2NDu)$$

so that

$$\|\chi_F Q'((U\phi)^2)\|_{Y,1/2}^{1/2} \le \|\phi\|_2((2M)^{1/2} + (32M^{3/2}\alpha^{-1}N^{-1/2})^{1/2}).$$

Thus we have

$$\|U\phi\|_X \le C\|\phi\|_2$$

where C is independent of N. In a similar fashion

$$\|V\phi\|_{X^*} \le C\|\phi\|_2.$$

Finally we introduce the measure ν_0 on $[N]$ defined by

$$\nu_0\{k\} = \int_{E_k\cap G} f\, d\lambda.$$

Clearly

$$\nu_0(H) \le \|U\chi_H\|_X\|V\chi_H\|_{X^*} \le C^2\frac{|H|}{N}.$$

On the other hand,

$$\nu_0([N]) = \int_G f\, d\lambda \ge M^{-1}\int_G \theta f\, d\mu_1 \ge \frac{\alpha}{8M}.$$

It follows that if $H_0 = \{k \in [N] : \nu_0\{k\} \ge (16MN)^{-1}\alpha\}$ then $|H_0| \ge \alpha N/16MC^2$. We are now in a position to apply Proposition 2.3 with $a(s) = g(s)\chi_G(s)$, $b(s) = h(s)\chi_G(s)$ and with $\sigma = \tau$. Thus $ab = f\chi_G$ and if $H \subset H_0$ we have

$$\int_{\tau^{-1}H} f\chi_G d\lambda \le \frac{\alpha|H|}{16MN}.$$

By Proposition 2.3, $(L_2[N])_{H_0}$ is C'−lattice isomorphic to a C'−complemented sublattice of X where C' is independent of N. This demonstrates that condition (c) holds.■

7. Complemented subspaces of Banach lattices

We now give the first applications of the machinery developed in the preceding three sections.

THEOREM 7.1. *Let Y be a separable order-continuous Banach lattice which contains no complemented sublattice lattice-isomorphic to ℓ_2. Suppose X is an order-continuous Banach lattice which is isomorphic to a complemented subspace of Y. Then X is lattice-isomorphic to a complemented sublattice of $Y(L_2)$.*

PROOF: Suppose first that X is nonatomic. Then we can represent Y as a good order-continuous function space on some Polish probability space (K, μ) We can represent X as a good order-continuous function space on (Δ, λ) with the strong density property. According to Theorem 6.1 we can then find positive measure-continuous operators $P, Q, R :$ $C(\Delta) \to L_0(\mu)$ and a constant M so that for $f \in L_0(\lambda)_+$,

$$\int_\Delta f \, d\lambda \le \int_K Pf \, d\mu \le M \int_\Delta f \, d\lambda$$
$$\|Qf\|_{Y,1/2} \le M \|f\|_{X,1/2}$$
$$\|Rf\|_{Y^*,1/2} \le M \|f\|_{X^*,1/2}$$
$$Pf \le (Qf)^{1/2}(Rf)^{1/2}.$$

We may suppose that Q, R are P-continuous. Then we may find Borel functions k_Q, k_R on $K \times \Delta$ so that $\nu_s^Q = k_Q(s, .).\nu_s^P$ and $\nu_s^R = k_R(s, .).\nu_s^P$ almost everywhere on K. Let ρ be the measure on $K \times \Delta$ defined by

$$\rho(B) = \int_K \int_\Delta \chi_B(s, t) \, d\nu_s^P(t) \, d\mu(s).$$

We will then have $k_Q k_R \ge 1$ ρ−a.e. We define lattice homomorphisms $U, V : L_0(\lambda) \to$ $L_0(\rho)$ by

$$Uf(s, t) = k_Q^{1/2}(s, t)f(t)$$
$$Vf(s, t) = k_R^{1/2}(s, t)f(t).$$

Then:

$$\|Uf\|_{Y(L_2,\nu^P)} = \|(\int k_Q(s,t)|f(t)|^2\,d\nu_s^P(t))^{1/2}\|_Y$$

$$= \|\int k_Q(s,t)|f(t)|^2\,d\nu_s^P(t)\|_{Y,1/2}^{1/2}$$

$$= \|\int |f(t)|^2\,d\nu_s^Q(t)\|_{Y,1/2}^{1/2}$$

$$= \|Q(|f|^2)\|_{Y,1/2}^{1/2}$$

$$\le M^{1/2}\||f|^2\|_{X,1/2}^{1/2}$$

$$= M^{1/2}\|f\|_X.$$

Thus U maps X into $(Y(L_2;\nu^P))_{max}$.

Similarly we have $\|Vf\|_{Y^*(L_2;\nu^r)} \le M^{1/2}\|f\|_{X^*}$. Since $Y(L_2;\nu^P)^* = Y^*(L_2;\nu^P)$ we can apply Proposition 2.3 once we observe that for any Borel subset F of K,

$$\int_{K\times F} k_Q^{1/2}(s,t)k_R^{1/2}(s,t)\,d\rho(s,t) \ge \rho(K\times F)$$

$$= \int_K P\chi_F\,d\mu$$

$$\ge \lambda(F).$$

Thus by Proposition 2.3 X is isomorphic to a complemented sublattice of $Y(L_2,\nu^P)$ and hence of $Y(L_2)$ by Lemma 3.5.

To complete the proof in the general case, we can decompose X into atomic and nonatomic parts. It thus suffices to consider only the atomic part. In this case (Proposition 2.4) X lattice embeds as a complemented sublattice of $Y(\ell_2)$ which is itself isomorphic to a complemented sublattice of $Y(L_2)$ by Lemma 3.5.∎

THEOREM 7.2. *Let Y be a separable order-continuous Banach lattice which contains no complemented sublattice lattice-isomorphic to ℓ_2. Suppose X is a separable order-continuous Banach lattice which contains no complemented sublattice lattice-isomorphic to L_2. Suppose X is isomorphic to a complemented subspace of Y; then*

(a) *X is lattice-isomorphic to a complemented sublattice of $Y(\ell_2)$,*

(b) *There is a non-trivial band X_0 in X which is lattice-isomorphic to a complemented sublattice of Y.*

PROOF: It suffices to consider the case when X is nonatomic. We represent X and Y as in Theorem 7.1. In this case we can apply Theorem 6.3 to deduce that we may represent P, Q and R in the form:

$$Pf(s) = \sum_{n=1}^{\infty} a_n^P(s) f(\sigma_n s)$$

$$Qf(s) = \sum_{n=1}^{\infty} a_n^Q(s) f(\sigma_n s)$$

$$Rf(s) = \sum_{n=1}^{\infty} a_n^R(s) f(\sigma_n s)$$

where a_n^P, a_n^Q, a_n^R are nonnegative Borel functions on K satisfying $a_n^Q a_n^R \geq (a_n^P)^2$ a.e. and $\sigma_n : K \to \Delta$ are Borel maps, satisfying $\sigma_m(s) \neq \sigma_n(s)$ whenever $m \neq n$.

Denote by γ counting measure on $\mathbf{N}$. We define $U, V : L_0(\lambda) \to L_0(K \times \mathbf{N}, \mu \times \gamma)$ by

$$Uf(s,n) = (a_n^Q(s))^{1/2} f(\sigma_n s)$$

$$Vf(s,n) = (a_n^R(s))^{1/2} f(\sigma_n s).$$

Then

$$\|Uf\|_{Y(\ell_2)} = \|(\sum_{n=1}^{\infty} a_n^Q |f \circ \sigma_n|^2)^{1/2}\|_Y$$

$$= \|Q(|f|^2)\|_{Y,1/2}^{1/2}$$

$$\leq M^{1/2}\|f\|_X.$$

Similarly

$$\|Vf\|_{Y^*(\ell_2)} \leq M^{1/2}\|f\|_{X^*}.$$

If $F \subset \Delta$ is a Borel set then:

$$\sum_{n=1}^{\infty} \int_{\sigma_n^{-1}F} (a_n^Q(s) a_n^R(s))^{1/2} d\mu(s) \geq \sum_{n=1}^{\infty} \int_{\sigma_n^{-1}F} a_n^P(s) \, d\mu(s)$$

$$= \int_K P\chi_F \, d\mu$$

$$\geq \lambda(F).$$

Thus by Proposition 2.3 X is lattice-isomorphic to a complemented sublattice of $Y(\ell_2)$, proving (a).

For (b) fix some n such that $a_n^P(s) > 0$ on a set of positive measure. Consider the lattice homomorphisms $U_n, V_n : L_0(\lambda) \to L_0(\mu)$ defined by:

$$U_n f(s) = (a_n^Q(s))^{1/2} f(\sigma_n s)$$

$$V_n f(s) = (a_n^R(s))^{1/2} f(\sigma_n s).$$

Then $\|U_n\|_{X \to Y_{max}}, \|V_n\|_{X^* \to Y^*} \leq M^{1/2}$. Further if F is a Borel subset of Δ, we set

$$\pi(F) = \int_{\sigma_n^{-1} F} a_n^P \, d\mu.$$

Thus π is a nontrivial λ−continuous measure on K_1. Hence there exists a Borel set E with $\lambda(E) > 0$ and $\delta > 0$ so that if $F \subset E$ then $\pi(F) \geq \delta \lambda(F)$. Thus if $F \subset E$ is an arbitrary Borel set,

$$\int_{\sigma_n^{-1} F} (a_n^Q a_n^R)^{1/2} d\mu \geq \pi(F) \geq \delta \lambda(F),$$

and we may apply Proposition 2.3 to show that X_E is lattice-isomorphic to a complemented sublattice of Y.∎

We now turn to the case of r.i. function spaces. Here Theorem 6.1 yields more.

THEOREM 7.3. Let Y be an order-continuous Banach lattice and suppose that X is a separable order-continuous rearrangement invariant function space on $[0,1]$, which is not equal to $L_2[0,1]$. Suppose X is isomorphic to a complemented subspace of Y. Then either Y contains a complemented sublattice lattice-isomorphic to the Haar representation of X or Y contains a complemented sublattice lattice-isomorphic to X. In either case Y contains a complemented sublattice isomorphic to X.

In particular, if the Boyd indices of X satisfy either $p_X = 1$ or $q_X = \infty$ then Y contains a complemented sublattice lattice-isomorphic to X.

PROOF: We first prove the theorem under the assumption that Y is separable. We may as in Theorem 7.1 represent Y as a function space on (K, μ), and we may suppose that X is modelled on (Δ, λ). Suppose that Y does not contain a complemented sublattice lattice-isomorphic to the Haar representation of X. Then we are in case (b) of Theorem 6.1 and since $X \neq L_2$ we are in case (b) of Theorem 6.3. We conclude exactly as in Theorem 7.2 above that there is a band in X lattice-isomorphic to a complemented sublattice of Y.

Since every band in X is lattice-isomorphic to X the theorem follows. The hypotheses on the Boyd indices in the last statement imply that the Haar basis fails to be unconditional.

We now consider the general case. The reduction to the previous case is attained by the observation that any separable subspace of Y is contained in a complemented separable sublattice. This follows from Proposition 1.a.3 (p. 9) of [33] combined with the complementation results of Amir and Lindenstrauss [4] on weakly compactly generated subspaces. More precisely we can first suppose that Y has a weak order-unit $u \geq 0$. If $Z = Z_0$ is a separable subspace of Y we can, by [4], construct a sequence of separable subspaces Z_n so that Z_{2n+1} is the smallest closed sublattice generated by Z_{2n} for $n \geq 0$ and Z_{2n} is a subspace containing Z_{2n-1} for $n \geq 1$ so that there is a projection $P_n : Y \to Z_{2n}$ with $\|P_n\| \leq 1$ and $P_n[-u, u] \subset [-u, u]$. Then (P_n) has a cluster-point P for the weak-operator topology which is a projection onto the closure of $\cup Z_{2n+1}$.∎

Now let us consider a special case of the above when Y is itself an r.i. space.

THEOREM 7.4. *Let X and Y be separable r.i. spaces on $[0,1]$. Suppose that X is not equal to $L_2[0,1]$ and that X is isomorphic to a complemented subspace of Y. Suppose that either (a) $p_X = 1$ or (b) $q_X = \infty$ or (c) Y contains no complemented sublattice lattice-isomorphic to the Haar representation H_X of X. Then $X = Y$.*

PROOF: It follows from Theorem 7.3 that in each case, Y contains a complemented sub-lattice lattice-isomorphic to X. However Proposition 2.5 then immediately yields $X = Y$.∎

The following theorem is an improvement of Theorems 5.1 and 6.1 in [22] in two ways. First we insist in the exceptional case that the Haar representation of X be isomorphic to a *complemented* sublattice of X. Second, we remove all side conditions on X; we do not assume either that the Haar basis is unconditional or that $X \supset L_p[0,1]$ for some $1 < p < 2$.

THEOREM 7.5. *Let X be a separable r.i. space on $[0,1]$. Suppose that either (a) $p_X = 1$ or (b) $q_X = \infty$ or (c) X contains no complemented sublattice lattice-isomorphic to the Haar representation of X. If Y is a separable r.i. space on $[0,1]$ such that X and Y are isomorphic then $X = Y$.*

PROOF: Cases (a) and (b) are covered by the preceding theorem. For (c) observe that we can assume that $1 < p_X \leq q_X < \infty$ and $1 < p_Y \leq q_Y < \infty$ so that both X and Y have Haar representations. Further $X \neq L_2$ and so $Y \neq L_2$. Thus if $X \neq Y$ we obtain that

H_Y is lattice-isomorphic to a complemented sublattice of X. Now X is isomorphic to H_Y and cannot be lattice isomorphic to a sublattice since H_Y is atomic. Thus by Theorem 7.3 H_Y has a complemented sublattice lattice-isomorphic to H_X. But this means that H_X is isomorphic to a complemented sublattice of X. ∎

THEOREM 7.6. *Suppose that Y is a separable r.i. space on $[0,1]$ and that X is a nonatomic Banach lattice which is isomorphic to a complemented subspace of Y. Suppose that Y contains no complemented sublattice which is lattice-isomorphic to ℓ_2 and that X contains no complemented sublattice which is lattice-isomorphic to L_2. Then X contains a nontrivial band which is lattice-isomorphic to Y.*

PROOF: This is immediate from Theorem 7.2 (b) and Proposition 2.5. ∎

Notice it then follows by an exhaustion argument that X can be decomposed as the direct sum of at most countably bands each lattice-isomorphic to Y. This suggests a uniqueness principle for lattice structures on Y which we investigate further in the next section for certain special cases. Notice that if Y is super-reflexive then Y is isomorphic to $Y(L_2)$ which explains the restriction on X. Finally we notice that X must contain a complemented copy of Y; in view of Proposition 2.d.5 (p. 172) of [33] this leads to:

COROLLARY 7.7. *Under the hypotheses of Theorem 7.6, suppose further that $1 < p_X \leq q_X < \infty$. Then X is isomorphic to Y.*

8. Strictly 2-concave and strictly 2-convex structures

We shall say that a Banach lattice X is strictly 2-concave if it is q-concave for some q where $1 < q < 2$ and is strictly 2-convex if it is p-convex for some p where $2 < p < \infty$. Our main theorem of this section is an extension of an implicit result of [22] which asserts that any nonatomic Banach lattice which is strictly 2-concave and isomorphic to a complemented subspace of L_p $(1 < p < 2)$ is lattice-isomorphic to L_p. However our proof is quite different.

THEOREM 8.1. *Suppose X is a strictly 2-concave r.i. space on $[0,1]$, with Boyd indices $\{p_X, q_X\}$, for which one of the following conditions holds:*

(a) $p_X = 1$

(b) $p_X < q_X$

(c) X is p_X-convex.

Let Y be a strictly 2-concave nonatomic Banach lattice. If Y is isomorphic to a complemented subspace of X then Y is lattice-isomorphic to X.

REMARK: If we assume that Y is isomorphic to X then we can relax the assumption that X is strictly 2-concave. Indeed, X must be of cotype 2 and cannot be equal to L_2; since Y is strictly 2-concave it cannot contain a sublattice, lattice-isomorphic to ℓ_2. By Theorem 7.3 X will embed into Y as a sublattice and hence will be strictly 2-concave.

We also note that (c) includes (a), but as we shall see the proof of (a) is rather simpler and it is this case which will be exploited in later results.

Before proving Theorem 8.1 we note the following dual result which can be deduced immediately or proved in the same way.

THEOREM 8.2. *Suppose X is a separable strictly 2-convex r.i. space on $[0,1]$, satisfying one of the following three conditions:*

(a) $q_X = \infty$

(b) $p_X < q_X$

(c) X is q_X-concave.

Let Y be a strictly 2-convex nonatomic order-continuous Banach lattice. If Y is isomorphic to a complemented subspace of X then Y is lattice-isomorphic to X.

PROOF OF THEOREM 8.1: The proof of Theorem 8.1 is somewhat long and involves several preliminary lemmas. We start by introducing an r.i. space on $[0, \infty)$; we define Z to be the space (cf. [22] or [33] p.203) of all $f \in L_0[0, \infty)$ such that $f^* \chi_{[0,1]} \in X$ and $f^* \chi_{[1,\infty)} \in L_2$. Z can be given a norm equivalent to the quasinorm $\|f^* \chi_{[0,1]}\|_X + \|f^* \chi_{[1,\infty)}\|_2$. Since X is 2-concave, the dual Z^* can be equivalently normed by

$$\|f\|_{Z^*} = \max\{\|f^* \chi_{[0,1]}\|_{X^*}, \|f\|_2\}$$

and Z^* is 2-convex. The norm on Z will be assumed to be the predual norm for this norm (which is plainly a dual norm).

We can and do suppose that Y is a good order-continuous function space on (Δ, λ), with the strong density property.

We next introduce the notion of a *special embedding* of Y^* into Z^*. Let $\Phi = (\phi_k)_{k \in \mathbf{Z}}$ be a doubly infinite sequence of nonnegative functions in $L_1(\Delta, \lambda)$. Define a σ-finite Borel measure μ on $\mathbf{Z} \times \Delta$ by

$$\mu(\{k\} \times E) = \int_E \phi_k \, d\lambda$$

for any Borel subset E of Δ. Then consider the map $T = T_\Phi : L_0(\Delta, \lambda) \to L_0(\mathbf{Z} \times \Delta, \mu)$ defined by

$$Tf(k, s) = 2^k f(s).$$

T is plainly a lattice homomorphism. We shall say that T is a *special embedding* if for some C and all $f \geq 0$,

$$\frac{1}{C}\|f\|_{Y^*} \leq \|Tf\|_{Z^*} \leq C\|f\|_{Y^*}.$$

In this case it quickly follows that we must have $\sum_{k \in \mathbf{Z}} \phi_k(s) > 0$ a.e. Define Borel functions u_Φ, v_Φ on Δ as follows:

$$u_\Phi(s) = \sup\{k : \phi_k(s) > 0\}$$
$$v_\Phi(s) = \inf\{k : \phi_k(s) > 0\}$$

so that $-\infty \leq v_\Phi \leq u_\Phi \leq \infty$ a.e.

LEMMA 8.3. *If $T = T_\Phi$ is a special embedding for which there exists $N \in \mathbf{N}$ such that $u_\Phi - v_\Phi \le N$ a.e., then Y is lattice-isomorphic to X.*

PROOF: Let $\phi(s) = \sum_{k \in \mathbf{Z}} \phi_k(s)$; the hypotheses imply that $\phi < \infty$ a.e. and hence there is a σ-finite measure ν on Δ with $d\nu = \phi d\lambda$. Then for $f \ge 0$,

$$\|Tf\|_{Z^*} \le \|2^{u_\Phi} f\|_{Z^*(\Delta,\nu)} \le 2^N \|Tf\|_{Z^*}.$$

Hence $\|f\|_{Y^*}$ is equivalent to $\|2^{u_\Phi} f\|_{Z^*(\Delta,\nu)}$ and Y^* is lattice isomorphic to $Z^*[0,\alpha)$ where $\alpha = \int \phi \, d\lambda$. Clearly since Y^* is strictly 2-convex, we must have $\alpha < \infty$. Furthermore $\|f\|_Y$ is equivalent to $\|2^{-u_\Phi} f\|_{Z(\Delta,\nu)}$ and so Y is lattice-isomorphic to $Z[0,\alpha)$ and hence to X. ∎

LEMMA 8.4. *Suppose $p_X = 1$ and that T_Φ and T_Ψ are two special embeddings of Y^* into Z^*. Suppose that $u_\Phi, v_\Phi, u_\Psi, v_\Psi$ are all finite a.e. Then there exists $N \in \mathbf{N}$ such that $|u_\Phi - u_\Psi| \le N$ a.e.*

PROOF: We suppose C is a constant such that

$$\frac{1}{C}\|f\|_{Y^*} \le \|T_\Phi f\|_{Z^*}, \|T_\Psi f\|_{Z^*} \le C\|f\|_{Y^*},$$

for $f \in L_0(\Delta, \lambda)$. It will suffice to prove a uniform estimate on $|u_\Phi - u_\Psi|$ on any set E of positive measure such that u_Φ, v_Φ, u_Ψ and v_Ψ are constant.

For $k \in \mathbf{Z}$, define

$$\alpha_k = \int_E \phi_k \, d\lambda, \qquad \beta_k = \int_E \psi_k \, d\lambda.$$

Then $(\alpha_k)_{k \in \mathbf{Z}}$ and $(\beta_k)_{k \in \mathbf{Z}}$ are each finitely non-zero. Let δ be the minimum of all positive values of α_k, β_k. Then define

$$\rho = \delta^{-1}\left(\sum_{k \in \mathbf{Z}}(\alpha_k + \beta_k)\right).$$

We use the fact that $q_{X^*} = \infty$. We can find a simple function $h \in X^*[0,1]_+$ of the form

$$h = \sum_{j=1}^{n} b_j \chi_{[(j-1)\eta, j\eta)}$$

so that $n\eta < \min(1/\rho, \delta)$ and so that $\|h\|_{X^*} = 1$ but $\|D_\rho h\|_{X^*} \le 2$. (D_ρ denotes the usual dilation operator.)

By Lyapunov's theorem we can find n disjoint Borel subsets $F_1, \ldots, F_n$ of E so that

$$\int_{F_j} \phi_k \, d\lambda = \delta^{-1} \eta \alpha_k$$

$$\int_{F_j} \psi_k \, d\lambda = \delta^{-1} \eta \beta_k$$

for $1 \leq j \leq n$ and $k \in \mathbf{Z}$. Then let $g = \sum_{j=1}^n b_j \chi_{F_j}$.

$$T_\Phi g = \sum_{k \in \mathbf{Z}} \sum_{j=1}^n b_j 2^k \chi_{\{k\} \times F_j}$$

$$\leq 2^{u_\Phi} \sum_{j=1}^n b_j \sum_{k \in \mathbf{Z}} \chi_{\{k\} \times F_j}.$$

Hence recalling that u_Φ is constant on E and denoting its constant value by u_Φ,

$$\|T_\Phi g\|_{Z^*} \leq 2^{u_\Phi} \|D_{(\delta^{-1} \sum \alpha_k)} h\|_{X^*}$$

$$\leq 2^{u_\Phi} \|D_\rho h\|_{X^*}$$

$$\leq 2^{u_\Phi + 1}.$$

Similarly

$$T_\Phi g \geq 2^{u_\Phi} \sum_{j=1}^n b_j \chi_{\{u_\Phi\} \times F_j}$$

and hence

$$\|T_\Phi g\|_{Z^*} \geq 2^{u_\Phi}.$$

Combining with similar estimates for T_Ψ we obtain:

$$C^{-1} 2^{u_\Phi} \leq C 2^{u_\Psi + 1}$$

$$C^{-1} 2^{u_\Psi} \leq C 2^{u_\Phi + 1}$$

so that

$$|u_\Phi - u_\Psi| \leq \log_2(2C^2) \blacksquare$$

LEMMA 8.5. *Let* $r = p_{X^*}$ *or* $r = q_{X^*}$, *and that* $r < \infty$. *Suppose* T_Φ *and* T_Ψ *are two special embeddings such that* $u_\Phi, v_\Phi, u_\Psi, v_\Psi$ *are finite a.e. Then there is a constant* L *so that*

$$L^{-1} \sum_{k \in \mathbf{Z}} 2^{kr}\phi_k \leq \sum_{k \in \mathbf{Z}} 2^{kr}\psi_k \leq L \sum_{k \in \mathbf{Z}} 2^{kr}\phi_k \quad a.e.$$

PROOF: The initial steps are similar to those of Lemma 8.4. We suppose that C is a constant such that

$$C^{-1}\|f\|_{Y^*} \leq \|T_\Phi f\|_{Z^*}, \|T_\Psi f\|_{Z^*} \leq C\|f\|_{Z^*}$$

for all $f \in L_0(\lambda)$, and that E is a Borel subset of Δ of positive measure so that the functions $u_\Phi, v_\Phi, u_\Psi, v_\Psi$ are all constant on E. For $k \in \mathbf{Z}$, define

$$\alpha_k = \int_E \phi_k \, d\lambda, \qquad \beta_k = \int \psi_k \, d\lambda.$$

Let $\delta > 0$ be the minimum of all positive values of α_k, β_k. Let $\rho = \delta^{-1}(\sum(\alpha_k + \beta_k))$.

Let m be a natural number with $m > 2\rho$. Then by Theorem 2.b.6 (p. 141) of [33] there exists $\eta > 0$, $n \in \mathbf{N}$ with $mn\eta < \min(1, \sum(\alpha_k + \beta_k))$ and a function $h \in X^*[0,1]_+$ with $\|h\|_{X^*} = 1$, of the form

$$h = \sum_{j=1}^n b_j \chi_{[(j-1)\eta, j\eta)}$$

so that if h_l, $(1 \leq l \leq m)$ are defined by

$$h_l(t) = h(t - (l-1)n\eta) \qquad (l-1)n\eta \leq t < ln\eta$$
$$h_l(t) = 0 \qquad \qquad \text{otherwise}$$

then, for all scalars $(\xi_l)_{l=1}^m$,

$$\frac{1}{2}(\sum_{l=1}^m |\xi_l|^r)^{1/r} \leq \|\sum_{l=1}^m \xi_l h_l\|_{X^*} \leq 2(\sum_{l=1}^m |\xi_l|^r)^{1/r}.$$

As in Lemma 8.4 we apply Lyapunov's theorem to find n disjoint Borel subsets $F_1, \ldots, F_n$ of E so that for $1 \leq j \leq n$ and $k \in \mathbf{Z}$,

$$\int_{F_j} \phi_k \, d\lambda = \delta^{-1}\eta\alpha_k$$
$$\int_{F_j} \psi_k \, d\lambda = \delta^{-1}\eta\beta_k.$$

We again let $g = \sum_{j=1}^{n} b_j \chi_{F_j}$. Then

$$T_\Phi g = \sum_{k \in \mathbf{Z}} 2^k \chi_{G_{jk}}$$

where $\mu(G_{jk}) = \delta^{-1} \eta \alpha_k$. (Here $\mu = \mu(\Phi)$.)

Now $\delta^{-1} \alpha_k \geq 1$ and so if γ_k is defined to be the integer part of $\delta^{-1} \alpha_k$ then $\gamma_k \leq \delta^{-1} \alpha_k \leq 2\gamma_k$. We can then partition G_{jk} into sets G_{ijk} for $1 \leq i \leq \gamma_k + 1$ where $\mu(G_{ijk}) = \eta$ for $i \leq \gamma_k$ and $\mu(G_{ijk}) < \eta$ for $i = \gamma_k + 1$.

Thus

$$T_\Phi g = \sum_{j=1}^{n} b_j \sum_{k \in \mathbf{Z}} 2^k \sum_{i=1}^{\gamma_k + 1} \chi_{G_{ijk}}$$
$$= \sum_{k \in \mathbf{Z}} 2^k \sum_{i=1}^{\gamma_k + 1} \left(\sum_{j=1}^{n} b_j \chi_{G_{ijk}} \right)$$

Note that

$$\sum_{k \in \mathbf{Z}} (\gamma_k + 1) \leq 2 \sum_{k \in \mathbf{Z}} \delta^{-1} \alpha_k \leq 2\rho < m.$$

Thus

$$\frac{1}{2} \left(\sum_{k \in \mathbf{Z}} 2^{kr} \gamma_k \right)^{1/r} \leq \| T_\Phi g \|_{Z^*} \leq 2 \left(\sum_{k \in \mathbf{Z}} 2^{kr} (\gamma_k + 1) \right)^{1/r}.$$

Since $\frac{1}{2} \delta^{-1} \alpha_k \leq \gamma_k < \gamma_k + 1 \leq 2\delta^{-1} \alpha_k$, we obtain

$$2^{-(r+1)} \delta^{-1} \sum_{k \in \mathbf{Z}} 2^{kr} \alpha_k \leq \| T_\Phi g \|_{Z^*}^r \leq 2^{r+1} \delta^{-1} \sum_{k \in \mathbf{Z}} 2^{kr} \alpha_k.$$

Similarly

$$2^{-(r+1)} \delta^{-1} \sum_{k \in \mathbf{Z}} 2^{kr} \beta_k \leq \| T_\Psi g \|_{Z^*}^r \leq 2^{r+1} \delta^{-1} \sum_{k \in \mathbf{Z}} 2^{kr} \beta_k.$$

Thus if we let $L = C^2 2^{2(r+1)}$ then

$$L^{-1} \sum_{k \in \mathbf{Z}} 2^{kr} \alpha_k \leq \sum_{k \in \mathbf{Z}} 2^{kr} \beta_k \leq L \sum_{k \in \mathbf{Z}} 2^{kr} \alpha_k.$$

We now recall that E was an arbitrary set of positive measure on which the functions u_Φ, v_Φ, u_Ψ and v_Ψ are constant. Thus for any such set

$$L^{-1} \int_E \left(\sum_{k \in \mathbf{Z}} 2^{kr} \phi_k \right) d\lambda \leq \int_E \left(\sum_{k \in \mathbf{Z}} 2^{kr} \psi_k \right) d\lambda \leq L \int_E \left(\sum_{k \in \mathbf{Z}} 2^{kr} \phi_k \right) d\lambda$$

and it follows that, a.e.,

$$L^{-1} \sum_{k \in \mathbf{Z}} 2^{kr} \phi_k \le \sum_{k \in \mathbf{Z}} 2^{kr} \psi_k \le L \sum_{k \in \mathbf{Z}} 2^{kr} \phi_k. \blacksquare$$

We now return to the proof of Theorem 8.1. We regard X as being modelled on (Δ, λ). Since X and Y are both strictly 2-concave, in particular neither contains any sublattice lattice-isomorphic to ℓ_2 or L_2. Thus Theorems 6.1 and 6.3 are applicable. We deduce the existence of a constant M, a sequence of Borel maps $\sigma_n : \Delta \to \Delta$ with $\sigma_m(s) \ne \sigma_n(s)$ if $m \ne n$ and sequences of nonnegative Borel functions a_n^P, a_n^Q, a_n^R on Δ so that $a_n^P \le (a_n^Q a_n^R)^{1/2}$ a.e. and if

$$Pf(s) = \sum_{n=1}^{\infty} a_n^P(s) f(\sigma_n s)$$

$$Qf(s) = \sum_{n=1}^{\infty} a_n^Q(s) f(\sigma_n s)$$

$$Rf(s) = \sum_{n=1}^{\infty} a_n^R(s) f(\sigma_n s)$$

then P, Q, R are positive measure-continuous operators from $C(\Delta)$ to $L_0(\lambda)$ satisfying

$$\int f \, d\lambda \le \int Pf \, d\lambda \le M \int f \, d\lambda$$

$$\|Qf\|_{X,1/2} \le M \|f\|_{Y,1/2}$$

$$\|Rf\|_{X^*,1/2} \le M \|f\|_{Y^*,1/2}$$

for all $f \in L_0(\lambda)_+$.

For $n \in \mathbf{N}$, $k \in \mathbf{Z}$, let $E(n,k) = \{s : 2^{2k} \le a_n^R(s) < 2^{2(k+1)}\}$. We will define two sequences of measures ω_k, ν_k on Δ by

$$\omega_k(F) = \sum_{n=1}^{\infty} \lambda(\sigma_n^{-1} F \cap E(n,k))$$

$$\nu_k(F) = \sum_{n=1}^{\infty} \int_{E(n,k) \cap \sigma_n^{-1} F} a_n^P(s) d\lambda(s).$$

Our first objective is to show that each of these measures is λ–continuous and to define its derivative. First observe that

$$\sum_{k \in \mathbf{Z}} \nu_k(F) = \int P\chi_F \, d\lambda \le M\lambda(F)$$

and so we can use the Radon-Nikodym Theorem to obtain nonnegative Borel functions $(w_k)_{k \in \mathbf{Z}}$ so that $d\nu_k = w_k d\lambda$ and $1 \leq \sum_{k \in \mathbf{Z}} w_k \leq M$ a.e.

In order to consider the measures ω_k we introduce the measure $\hat{\lambda}$ on $\mathbf{N} \times \Delta$ defined by $\hat{\lambda}(\{n\} \times E) = \lambda(E)$. We define $S : Y^* \to L_0(\mathbf{N} \times \Delta, \hat{\lambda})$ by

$$Sf(n,t) = (a_n^R(t))^{1/2} f(\sigma_n t).$$

Then Z^* is 2-convex and so, applying Lemma 7.2 of [22] to its 2-concavification, for a suitable constant C we have

$$\|Sf\|_{Z^*(\hat{\lambda})} \leq C \|(\sum_{n=1}^{\infty} a_n^R (f \circ \sigma_n)^2)^{1/2}\|_{X^*}$$
$$= C \|R(f^2)\|_{X^*,1/2}^{1/2}$$
$$\leq C M^{1/2} \|f\|_{Y^*}.$$

Thus $S : Y^* \to Z^*(\hat{\lambda})$ is bounded. As $S\chi_F \geq 2^k$ on a set measure of measure at least $\omega_k(F)$, we conclude that each ω_k is a finite λ-continuous measure and we introduce the nonnegative Borel functions $(\phi_k)_{k \in \mathbf{Z}}$ such that $d\omega_k = \phi_k d\lambda$.

The next step is to use the functions (w_k) to construct three lattice embeddings of Y^* into Z^*. We first determine four measurable integer-valued functions $(\kappa_j)_{j=1}^4$ on Δ so that if $\max_k w_k(s) \geq \frac{1}{8}$ then $\kappa_1(s) = \kappa_2(s) = \kappa_3(s) = \kappa_4(s) = l$ where $w_l(s) \geq \frac{1}{8}$; if, on the other hand, $\max_k w_k(s) < \frac{1}{8}$ then $\kappa_j(s)$ are such that:

$$\sum_{k \leq \kappa_1(s)} w_k(s) \leq \frac{1}{8}$$

$$\frac{1}{4} \leq \sum_{k \leq \kappa_2(s)} w_k(s) < \frac{3}{8}$$

$$\frac{1}{2} \leq \sum_{k \leq \kappa_3(s)} w_k(s) < \frac{5}{8}$$

$$\frac{3}{4} \leq \sum_{k \leq \kappa_4(s)} w_k(s) < \frac{7}{8}.$$

It follows that $\kappa_1 \leq \kappa_2 \leq \kappa_3 \leq \kappa_4$ and that for $j = 1, 2, 3$,

$$\sum_{k=\kappa_j(s)}^{\kappa_{j+1}(s)} w_k(s) \geq \frac{1}{8}.$$

We now introduce the Borel sets $V_j(k) = \{t : \kappa_j(t) \leq k \leq \kappa_{j+1}(t)\}$ for $j = 1, 2, 3$ and $k \in \mathbf{Z}$. For $n \in \mathbf{N}$ and $j = 1, 2, 3$ we define $W_j(n) = \cup_{k \in \mathbf{Z}}(E(n, k) \cap \sigma_n^{-1} V_j(k))$. Notice that this union is a disjoint union for each fixed n.

We now use these sets to cut up the original operators P, Q, R. We define for $j = 1, 2, 3$,

$$P_j f(t) = \sum_{n=1}^{\infty} a_n^{P_j}(t) f(\sigma_n t)$$

$$Q_j f(t) = \sum_{n=1}^{\infty} a_n^{Q_j}(t) f(\sigma_n t)$$

$$R_j f(t) = \sum_{n=1}^{\infty} a_n^{R_j}(t) f(\sigma_n t)$$

where $a_n^{P_j} = a_n^P \chi_{W_j(n)}$, $a_n^{Q_j} = a_n^Q \chi_{W_j(n)}$ and $a_n^{R_j} = a_n^R \chi_{W_j(n)}$. It is clear, with these definitions that P_j, Q_j, R_j are measure continuous operators on $C(\Delta)$ with $0 \leq P_j \leq P$, $0 \leq Q_j \leq Q$, $0 \leq R_j \leq R$ and $P_j \leq (Q_j R_j)^{1/2}$.

For any Borel set E we have:

$$\int P_j \chi_E \, d\lambda = \sum_{n=1}^{\infty} \int_{W_j(n)} a_n^P \chi_{\sigma_n^{-1} E} \, d\lambda$$

$$= \sum_{n=1}^{\infty} \sum_{k \in \mathbf{Z}} \int_{E(n,k)} a_n^P \chi_{\sigma_n^{-1}(V_j(k) \cap E)} \, d\lambda$$

$$= \sum_{k \in \mathbf{Z}} \nu_k(V_j(k) \cap E)$$

$$= \sum_{k \in \mathbf{Z}} \int_E w_k \chi_{V_j(k)} \, d\lambda$$

$$\geq \frac{1}{8} \lambda(E).$$

It follows that:

$$\int P_j f \, d\lambda \geq \frac{1}{8} \int f \, d\lambda$$

for $f \in L_1(\lambda)_+$.

Now since Y is strictly 2-concave, it does not contain sublattices uniformly lattice-isomorphic to ℓ_2^n for $n \in \mathbf{N}$. Thus Theorem 6.4 is also applicable and we can deduce the existence of $c > 0$ so that for $j = 1, 2, 3$ and for all $f \geq 0$ in $L_1(\lambda)_+$,

$$\int \sup_n (a_n^{P_j}(s) f(\sigma_n s)) \, d\lambda(s) \geq c \int f(s) \, d\lambda(s).$$

Now for $f \in Y^*$ and $g \in Y$,

$$c \int |fg| \, d\lambda \leq \int \sup_n |a_n^{P_j}(f \circ \sigma_n)(g \circ \sigma_n)| \, d\lambda$$

$$\leq \| \sup_n |(a_n^{R_j})^{1/2} f \circ \sigma_n| \|_{X^*} \| \sup_n |(a_n^{Q_j})^{1/2} g \circ \sigma_n| \|_X$$

$$\leq M^{1/2} \|g\|_Y \| \sup_n |(a_n^{R_j})^{1/2} f \circ \sigma_n| \|_{X^*}.$$

By maximizing over $\|g\|_Y \leq 1$ we clearly have, for each $f \in Y^*$,

$$\| \sup_n |(a_n^{R_j})^{1/2} f \circ \sigma_n| \|_{X^*} \geq \frac{c}{M^{1/2}} \|f\|_{Y^*}.$$

We now use these results to cut S in a similar way. For $j = 1, 2, 3$ define $S_j : Y^* \to L_0(\mathbf{N} \times \Delta, \hat{\lambda})$ by

$$S_j f(n, t) = (a_n^{R_j}(t))^{1/2} f(\sigma_n t).$$

Since $0 \leq S_j \leq S$ each S_j is bounded from Y^* to Z^*. On the other hand

$$\|S_j f\|_{Z^*} \geq \| \sup_n |(a_n^{R_j})^{1/2} f \circ \sigma_n| \|_{X^*} \geq \frac{c}{M^{1/2}} \|f\|_{Y^*}.$$

Now we introduce three doubly infinite sequences of nonnegative $L_1(\lambda)$ functions by defining $\phi_k^{(j)} = \phi_k \chi_{V_j(k)}$. Let $\Phi^{(j)} = (\phi_k^{(j)})_{k \in \mathbf{Z}}$. We then create three measures $\{\mu_j : j = 1, 2, 3\}$ on $\mathbf{Z} \times \Delta$ by setting

$$\mu_j(\{k\} \times E) = \int_E \phi_k^{(j)} d\lambda.$$

Consider the maps $T_j = T_{\Phi^{(j)}} : Y^* \to L_0(\mathbf{Z} \times \Delta)$ defined by

$$T_j f(k, t) = 2^k f(t).$$

We will show that each T_j is an embedding of Y^* into $Z^*(\mu_j)$ and hence is a special embedding. Suppose $f \geq 0$ in Y^*. Set $G(\tau) = \{s : f(s) > \tau\}$. Then

$$\hat{\lambda}(|S_j f| > \tau) = \sum_{n=1}^{\infty} \lambda((a_n^{R_j})^{1/2} f \circ \sigma_n > \tau)$$

$$= \sum_{n=1}^{\infty} \sum_{k \in \mathbf{Z}} \lambda(((a_n^R)^{1/2} f \circ \sigma_n > \tau) \cap \sigma_n^{-1} V_j(k) \cap E(n, k)).$$

Thus

$$\hat{\lambda}(|S_j f| > \tau) \geq \sum_{n=1}^{\infty} \sum_{k \in \mathbf{Z}} \lambda(\sigma_n^{-1}(G(\tau 2^{-k}) \cap V_j(k)) \cap E(n, k))$$

$$= \sum_{k \in \mathbf{Z}} \omega_k(G(\tau 2^{-k}) \cap V_j(k))$$

$$= \sum_{k \in \mathbf{Z}} \int_{G(\tau 2^{-k})} \phi_k^{(j)} d\lambda$$

$$= \mu_j(|T_j f| > \tau).$$

Similarly $\hat{\lambda}(|S_j f| > \tau) \leq \mu_j(|T_j f| \geq \tau/2)$. Combining these we see that each $T_j : Y^* \to Z^*(\mu_j)$ is a special embedding. Furthermore if $u_j = u_{\Phi(j)}$ and $v_j = v_{\Phi(j)}$ then we have $\kappa_j \leq v_j \leq u_j \leq \kappa_{j+1}$ a.e.

We are now in position to apply Lemmas 8.3-8.5.

Case (a): It follows from Lemma 8.4 that $u_3 - u_1$ is essentially bounded and hence so is $u_2 - v_2$. Now Lemma 8.3 completes the proof.

Case (b): We can assume $p_X > 1$. Let $p = p_{X^*}$ and $q = q_{X^*}$ so that $p < q$. Then for a suitable constant L we have a.e.

$$L^{-1} \sum_{k \in \mathbf{Z}} 2^{kr} \phi_k^{(1)} \leq \sum_{k \in \mathbf{Z}} 2^{kr} \phi_k^{(3)} \leq L \sum_{k \in \mathbf{Z}} 2^{kr} \phi_k^{(1)}$$

where $r = p$ or $r = q$ by Lemma 8.5. Now, almost everywhere,

$$\sum_{k \in \mathbf{Z}} 2^{kq} \phi_k^{(3)} \geq 2^{(q-p)v_3} \sum_{k \in \mathbf{Z}} 2^{kp} \phi_k^{(3)}$$

$$\geq L^{-1} 2^{(q-p)v_3} \sum_{k \in \mathbf{Z}} 2^{kp} \phi_k^{(1)}$$

$$\geq L^{-1} 2^{(q-p)(v_3 - u_1)} \sum_{k \in \mathbf{Z}} 2^{kq} \phi_k^{(1)}$$

$$\geq L^{-2} 2^{(q-p)(v_3 - u_1)} \sum_{k \in \mathbf{Z}} 2^{kq} \phi_k^{(3)}$$

so that $v_3 - u_1$ is essentially bounded. Hence $u_2 - v_2$ is essentially bounded and we complete the proof by Lemma 8.3.

Case (c): Here X is p-convex where $p = p_X$; we can assume that the p-convexity constant is one, and the same will then hold for Z. We let $q = q_{X^*}$ be the conjugate index in this case. We will introduce measures $\hat{\mu}_1$ and $\hat{\mu}_3$ on $\mathbf{Z} \times \Delta$ by setting

$$\hat{\mu}_j(\{k\} \times E) = \int_E 2^{q(k-\kappa_2)} \phi_k^{(j)} d\lambda.$$

Now define $\hat{T} : L_0(\lambda) \to L_0(\hat{\mu}_1 + \hat{\mu}_3)$ by $\hat{T}f(k,t) = 2^{\kappa_2(t)} f(t)$. If we can show that $\hat{T}$ is an embedding of Y^* into $Z^*(\hat{\mu}_1 + \hat{\mu}_3)$ then it will follow that Y is lattice-isomorphic to X by an argument similar to the one used in the proof of Lemma 8.3.

First we notice that Lemma 8.5 implies that there is a constant C so that

$$C^{-1}\|\hat{T}f\|_{Z^*(\hat{\mu}_1)} \leq \|\hat{T}f\|_{Z^*(\hat{\mu}_3)} \leq C\|\hat{T}f\|_{Z^*(\hat{\mu}_1)}.$$

Now suppose $h \in Z(\hat{\mu}_1)$ with $\|h\|_Z = 1$. Then

$$\int h(\hat{T}f)d\hat{\mu}_1 = \sum_{k \in \mathbf{Z}} \int h(k,t)2^{\kappa_2(t)+q(k-\kappa_2(t))} f(t)\phi_k^{(1)}(t)d\lambda(t)$$

$$= \int_{\mathbf{Z} \times \Delta} g(k,t)T_1 f(k,t)d\mu_1(t)$$

where

$$g(k,t) = 2^{(q-1)(k-\kappa_2(t))} h(k,t).$$

Thus

$$|g|^p = U(|h|^p)$$

where

$$UF(k,t) = 2^{q(k-\kappa_2)} F(k,t).$$

It is easily seen that U is an isometry from $L_1(\hat{\mu}_1)$ into $L_1(\mu_1)$ which is norm-decreasing for the L_∞−norm. Since Z is p-convex with constant one we conclude that $\|g\|_{Z(\mu_1)} \leq 1$ and hence that

$$\int h(\hat{T}f)d\hat{\mu}_1 \leq \|T_1 f\|_{Z^*(\mu_1)}$$

whence we conclude that

$$\|\hat{T}f\|_{Z^*(\hat{\mu}_1)} \leq \|T_1 f\|_{Z^*(\mu_1)}.$$

For the converse direction suppose now that $h \in Z(\mu_3)$ with $\|h\|_{Z(\mu_3)} = 1$. Then for any $f \in Z^*(\lambda)$,

$$\int h(T_3 f)\, d\mu_3 = \sum_{k \in \mathbf{Z}} \int h(k,t) 2^k f(t) \phi_k^{(3)}(t) d\lambda(t)$$

$$= \int_{\mathbf{Z} \times \Delta} h(k,t) 2^{k+q(\kappa_2(t)-k)} f(t) d\hat{\mu}_3(t)$$

$$= \int_{\mathbf{Z} \times \Delta} g(k,t) \hat{T} f(t) d\hat{\mu}_3(t)$$

where $g(k,t) = 2^{(q-1)(\kappa_2(t)-k)} h(k,t)$. Thus $|g|^p = U'(|h|^p)$ where

$$U' F(k,t) = 2^{q(\kappa_2(t)-k)} F(k,t).$$

As before U' is an isometry from $L_1(\mu_3)$ to $L_1(\hat{\mu}_3)$ which is also L_∞−norm decreasing. Thus $\|g\|_{Z(\mu_3)} \leq 1$ and so as before

$$\|\hat{T} f\|_{Z^*(\hat{\mu}_3)} \leq \|T_3 f\|_{Z^*(\mu_3)}.$$

Combining the above we conclude that $\hat{T}$ is an embedding of Y^* into $Z^*(\hat{\mu}_1 + \hat{\mu}_3)$ and the proof is complete.∎

9. Uniqueness of lattice structure

The following proposition is closely related to [6], Lemma 7.13.

PROPOSITION 9.1. *Let X be an order-continuous Banach lattice. Suppose X is 2-concave (resp. 2-convex); then X fails to be strictly 2-concave (resp. strictly 2-convex) if and only if there is a constant C so that for every $n \in \mathbf{N}$, X contains a C-complemented sublattice which is C-lattice-isomorphic to ℓ_2^n.*

PROOF: One direction is trivial since if ℓ_2^n is C-lattice isomorphic to a sublattice of X for every n then X cannot be either strictly 2-concave or strictly 2-convex. We will prove the other direction for the separable case when we may suppose that X is a Köthe function space on some (K, μ) where K is a Polish space and μ is a σ–finite Borel measure. (The non-separable case can be proved by using the separable complementation property, but we will not need it.) We will restrict attention to the case when X is 2-concave, but not strictly 2-concave; the other case has a similar proof. We can and do assume that the 2-concavity constant of X is one.

Now X^* is 2-convex with constant one which implies that $X_{1/2}^*$ is a Köthe function space. Let c_n be the least constant so that for disjoint $f_1, \ldots, f_n$ we have

$$\|f_1 + \cdots + f_n\|_{X^*} \leq c_n \max \|f_i\|_{X^*}.$$

Then $c_{mn} \leq c_m c_n$. Hence if for any n we have $c_n < n^{1/2}$ then there exists $\alpha < 1/2$ so that $c_n = O(n^\alpha)$. Pick p with $\alpha < 1/p < 1/2$. Then for any disjoint $f_1, f_2, \ldots, f_n$ with $\|f_i\|_{X^*}$

decreasing we have, setting $f_i = 0$ for $i > n$,

$$\|\sum_{i=1}^{n} f_i\|_{X^*} \leq \sum_{k=1}^{\infty} \| \sum_{i=2^{k-1}}^{2^k-1} f_i\|_{X^*}$$

$$\leq C \sum_{k=1}^{\infty} 2^{\alpha(k-1)}\|f_{2^{k-1}}\|_{X^*}$$

$$\leq C \sum_{k=1}^{\infty} 2^{(\alpha-1/p)(k-1)}(\sum_{i=1}^{n} \|f_i\|^p)^{1/p}$$

$$\leq C'(\sum_{i=1}^{n} \|f_i\|^p)^{1/p}$$

where C, C' are constants. Hence X^* has an upper p-estimate and hence X is r-concave for some $r < 2$. This contradicts our assumption and so $c_n = n^{1/2}$ for all n.

Now if $n \in \mathbf{N}$ and $0 < \delta < 1$ we pick disjoint positive norm-one $f_1, \ldots, f_n$ so that

$$\|f_1 + \cdots + f_n\|_{X^*} > n^{1/2}(1 - 1/(2\delta n)).$$

Then

$$\|f_1^2 + \cdots + f_n^2\|_{X^*,1/2} > n - \delta.$$

Now $X_{1/2}^*$ is a Köthe function space and hence there exists $v \geq 0$ in its Köthe dual Z say so that $\|v\|_Z = 1$ and

$$\int v(f_1^2 + \cdots f_n^2)d\mu > n - \delta.$$

Hence for all j we have

$$\int v f_j^2 d\mu > 1 - \delta.$$

Now $v f_j$ is in the Köthe dual of X^* i.e. X for $1 \leq j \leq n$ and it is immediate that

$(1 - \delta) < \|v f_j\|_X \leq 1$. For $g \in X$ we let $a_j = \int f_j g \, d\mu / \int v f_j^2 \, d\mu$, and then we have

$$(1 - \delta)(\sum_{j=1}^{n} a_j^2)^{1/2} \leq \|\sum_{j=1}^{n} a_j v f_j\|_X$$

$$= \sup_{\|h\|_{X^*} \leq 1} \int \sum_{j=1}^{n} a_j v f_j h \, d\mu$$

$$\leq \sup_{\|h\|_{X^*} \leq 1} \|\sum_{j=1}^{n} a_j f_j h\|_{X^*, 1/2}$$

$$\leq \|\sum_{j=1}^{n} a_j f_j\|_{X^*}$$

$$\leq (\sum_{j=1}^{n} a_j^2)^{1/2}$$

while

$$\sum_{j=1}^{n} a_j^2 \leq (1 - \delta)^{-1} \int \sum_{j=1}^{n} a_j f_j g \, d\mu$$

$$\leq (1 - \delta)^{-1} \|\sum_{j=1}^{n} a_j f_j\|_{X^*} \|g\|_X$$

$$\leq (1 - \delta)^{-1} (\sum_{j=1}^{n} a_j^2)^{1/2} \|g\|_X.$$

These calculations show that $v f_j$ span a sublattice $(1 - \delta)^{-1}$–isomorphic to ℓ_2^n and $(1 - \delta)^{-1}$–complemented.∎

We now turn to the question of finding a condition on a Banach lattice such that it does not contain uniformly complemented ℓ_2^n's. If X is a Banach lattice we define $d_n = d_n(X)$ to be the least constant such that for disjoint $f_1, \ldots, f_n \in X$ we have

$$\sum_{i=1}^{n} \|f_i\| \leq d_n \|\sum_{i=1}^{n} f_i\|$$

and $e_n = e_n(X)$ to be the least constant so that for disjoint $f_1, \ldots, f_n$ we have

$$\|\sum_{i=1}^{n} f_i\| \leq e_n \max_{1 \leq i \leq n} \|f_i\|.$$

PROPOSITION 9.2. *Let X be a separable order-continuous Banach lattice. Then $d_n(X) = e_n(X^*)$ and $d_n(X^*) = e_n(X)$. Furthermore if $f_1, \ldots, f_n$ are any nonnegative elements of X then:*

$$\sum_{i=1}^{n} \|f_i\|_X \leq d_n \|\sum_{i=1}^{n} f_i\|_X$$

$$\|\max_{1 \leq i \leq n} f_i\|_X \leq e_n \max_{1 \leq i \leq n} \|f_i\|_X.$$

PROOF: We may assume that X is a Köthe function space on (K, μ). First we prove that $d_n(X) = e_n(X^*)$. Suppose $g_1, \ldots, g_n \in X^*$ are disjoint, and nonnegative. For $\epsilon > 0$ pick $f \in X$ so that $f \geq 0$, $\|f\|_X = 1$ and

$$\int f(g_1 + \cdots + g_n)\, d\mu > \|g_1 + \cdots + g_n\|_{X^*} - \epsilon.$$

Let E_j be the support of g_j. Then

$$\|\sum_{j=1}^{n} g_j\|_{X^*} < \sum_{j=1}^{n} \int fg_j\, d\mu + \epsilon$$

$$\leq \sum_{j=1}^{n} \|f\chi_{E_j}\|_X \|g_j\|_{X^*} + \epsilon$$

$$\leq \max_{1 \leq j \leq n} \|g_j\|_{X^*} (\sum_{j=1}^{n} \|f\chi_{E_j}\|_X) + \epsilon$$

$$\leq d_n(X) \max_{1 \leq j \leq n} \|g_j\|_{X^*} + \epsilon.$$

Thus $e_n(X^*) \leq d_n(X)$. For the converse suppose $f_1, \ldots, f_n \geq 0$ have disjoint supports $E_1, \ldots, E_n$ in X. Choose $g_j \in X^*$ so that g_j has support in E_j, $g_j \geq 0$, $\|g_j\|_{X^*} = 1$ and

$$\int f_j g_j\, d\mu = \|f_j\|_X.$$

Then

$$\sum_{j=1}^{n} \|f_j\|_X = \sum_{j=1}^{n} \int f_j g_j\, d\mu$$

$$\leq \|\sum_{j=1}^{n} f_j\|_X \|\sum_{j=1}^{n} g_j\|_{X^*}$$

$$\leq e_n(X^*) \|\sum_{j=1}^{n} f_j\|_X$$

and so $e_n(X^*) = d_n(X)$. Clearly the same arguments give $d_n(X^*) = e_n(X)$. Of the remaining statements the second is quite obvious. It remains to prove the first. Suppose $f_j \geq 0$ in X. Pick as above $g_j \in X^*$ with $g_j \geq 0$ so that $\|g_j\|_{X^*} = 1$ and

$$\int f_j g_j \, d\mu = \|f_j\|_X.$$

Then

$$\sum_{j=1}^{n} \|f_j\|_X = \int \sum_{j=1}^{n} f_j g_j \, d\mu$$

$$\leq \int (\max_{1 \leq j \leq n} g_j) \sum_{j=1}^{n} f_j \, d\mu$$

$$\leq \| \max_{1 \leq j \leq n} g_j \|_{X^*} \| \sum_{j=1}^{n} f_j \|_X$$

$$\leq e_n(X^*) \| \sum_{j=1}^{n} f_j \|_X$$

and the conclusion follows. $\blacksquare$

LEMMA 9.3. *Let X be a Banach lattice for which* $\liminf d_n n^{-\alpha} = 0$. *Then X is q-concave where* $q = (1 - \alpha)^{-1}$.

PROOF: We clearly have that d_n is submultiplicative, i.e. $d_{mn} \leq d_m d_n$ for all m, n. Thus we conclude that there exists $\beta < \alpha$ and C so that $d_n \leq Cn^\beta$. Suppose $(1 - \beta)^{-1} < r < (1 - \alpha)^{-1}$ and suppose $f_j, 1 \leq j \leq n$, are disjoint in X. We suppose $\|f_j\|_X$ is decreasing. Then for $1 \leq k \leq n$,

$$k\|f_k\|_X \leq \sum_{j=1}^{k} \|f_j\|_X \leq d_k \| \sum_{j=1}^{n} f_j \|_X.$$

Thus

$$\sum_{k=1}^{n} \|f_k\|_X^r \leq (\sum_{k=1}^{n} \frac{d_k^r}{k^r}) \| \sum_{j=1}^{n} f_j \|_X^r.$$

Now $d_k/k \leq Ck^{\beta-1}$ and $r(1 - \beta) > 1$ so that we obtain that X has a lower r-estimate. Hence X is q-concave (see [33] p.85). $\blacksquare$

We now come to our main application.

THEOREM 9.4. *Let X be a Banach lattice for which* $\liminf d_n (\log n)^{-1/2} = 0$ *Then* X *does not contain uniformly complemented* ℓ_2^n's.

PROOF: We remark that X is q-concave for any $q > 1$ by Lemma 9.3. Let us assume that X contains uniformly complemented ℓ_2^n's. Let $\mathcal{U}$ be any free ultrafilter on $\mathbf{N}$ and form the ultraproduct $X_\mathcal{U}$. Then $X_\mathcal{U}$ is also in a natural way a Banach lattice and it is easy to verify that $d_n(X_\mathcal{U}) = d_n(X)$ for all n. Further $X_\mathcal{U}$ contains a complemented copy of ℓ_2. It follows that there is a separable closed sublattice Y of $X_\mathcal{U}$ which contains a complemented subspace isomorphic to ℓ_2. Furthermore $d_n(Y) \leq d_n(X)$ for all n.

We thus can identify Y as a good order-continuous Köthe function space on a probability space (K, μ). We may suppose that there exist sequences $f_n \in Y$ and $g_n \in Y^*$ so that: $\int f_i g_i d\mu = 1$ for all i, $\int f_i g_j d\mu = 0$ if $i \neq j$ and for some constant C and all $a_1, \ldots, a_n \in \mathbf{R}$,

$$C^{-1}(\sum_{j=1}^{n} |a_j|^2)^{1/2} \leq \| \sum_{j=1}^{n} a_j f_j \|_X, \quad \| \sum_{j=1}^{n} a_j g_j \|_{X^*} \leq C(\sum_{j=1}^{n} |a_j|^2)^{1/2}.$$

We argue first no subsequence of $f_n g_n$ can converge to zero in measure. For if it does then we may find a subsequence $\phi_k = f_{n_k}$, $\psi_k = g_{n_k}$ and a sequence of disjoint Borel sets E_k so that

$$\int_{E_k} |\phi_k \psi_k| d\mu \geq \frac{1}{2}.$$

Then $\| \phi_k \chi_{E_k} \|_Y \geq 1/(2C)$ and so, since Y is 2-concave, for a suitable constant C_0 we have, for all $k \in \mathbf{N}$,

$$k \leq 2C \sum_{i=1}^{k} \| \phi_k \chi_{E_k} \|_Y$$

$$\leq 2C d_k(Y) \| \sum_{i=1}^{k} \phi_k \chi_{E_k} \|_Y$$

$$\leq 2C d_k(Y) \| (\sum_{i=1}^{k} |\phi_k|^2)^{1/2} \|_Y$$

$$\leq 2C C_0 d_k(Y) \operatorname*{Ave}_{\epsilon_i = \pm 1} \| \sum_{i=1}^{k} \epsilon_i \phi_i \|_Y$$

$$\leq 2C^2 C_0 d_k(Y) k^{1/2}$$

and this leads to a contradiction. We conclude the existence of $\delta > 0$ and a sequence of Borel sets F_n with $\mu(F_n) \geq \delta$ so that $|f_n g_n| \geq \delta \chi_{F_n}$ for all n.

Next notice that $\|f_n\|_1 \leq \|f_n\|_Y \leq 2C$ so that $\{|f_n| > 4C/\delta\}$ has measure at most $\delta/2$. Thus we may find Borel subsets G_n of F_n with $\mu G_n > \delta/2$ so that $|g_n| \geq (\delta^2/4C)\chi_{G_n}$. Hence for all n we have

$$\|\sum_{i=1}^{n} |g_i|\|_{Y^*} \geq \frac{\delta^2}{8C}\|\sum_{i=1}^{n} \chi_{G_i}\|_1 \geq \frac{\delta^3}{16C}n.$$

Thus

$$\frac{\delta^3}{16C}n \leq \|\sum_{i=1}^{n} |g_i|\|_{Y^*}$$

$$= \|\max_{\epsilon_i=\pm 1} |\sum_{i=1}^{n} \epsilon_i g_i|\|_{Y^*}$$

$$\leq e_{2^n}(Y^*) \max_{\epsilon_i=\pm 1} \|\sum_{i=1}^{n} \epsilon_i g_i\|_{Y^*}$$

$$\leq C d_{2^n}(Y)n^{1/2}.$$

As $\liminf d_{2^n}(Y)n^{-1/2} = 0$ we again obtain a contradiction.∎

THEOREM 9.5. *Let X be a separable order continuous r.i. space on $[0,1]$ for which $\liminf d_n(X)(\log n)^{-1/2} = 0$. If Y is a nonatomic Banach lattice isomorphic to a complemented subspace of X then Y is lattice-isomorphic to X.*

PROOF: Observe X is q-concave for all $q > 1$ by Lemma 9.3. It then follows (see [33] p.132) that $p_X = q_X = 1$. Thus X and hence Y is of cotype 2. In particular Y is 2-concave. However, Y does not contain uniformly complemented ℓ_2^n's by Theorem 9.4. By Proposition 9.1, Y is strictly 2-concave. The conclusion follows by Theorem 8.1. ∎

By duality we obtain:

THEOREM 9.6. *Let X be a separable order-continuous r.i. space on $[0,1]$ such that $\liminf e_n(X)(\log n)^{-1/2} = 0$. If Y is a nonatomic order-continuous Banach lattice isomorphic to a complemented subspace of X then Y is lattice isomorphic to X.*

PROOF: Just observe that X^* verifies the assumptions of the previous theorem, and since Y is order-continuous it is determined as a lattice by Y^*; more precisely if we represent Y

as a good Köthe function space on (K, μ) then Y_{max} is the Köthe dual of Y^* and Y is the closure of the simple functions in Y_{max}. ∎

We now turn to examples. We consider certain Orlicz spaces considered in [22] (p. 235) as examples of r.i. spaces on $[0, 1]$ which are not isomorphic to r.i. spaces on $[0, \infty)$. Define F_α by

$$F_\alpha(t) = \begin{cases} t, & 0 \leq t \leq e; \\ t(\log t)^\alpha, & \text{otherwise} \end{cases}$$

For $0 < \alpha < 1/2$ the Orlicz spaces L_{F_α} are given as examples of r.i. spaces on $[0, 1]$ which are not isomorphic to r.i. spaces on $[0, \infty)$. An important ingredient of this result is that ℓ_2 is not isomorphic to a complemented subspace of L_{F_α} (see Lemma 8.16 of [22]). The next lemma shows that we can refine this to show that the same spaces do not contain uniformly complemented ℓ_2^n's.

LEMMA 9.7. *For the Orlicz space L_{F_α} we have there is a constant C so that $d_n \leq C(\log n)^\alpha$ for $n \geq 2$.*

PROOF: The dual space of L_{F_α} is (up to equivalent norm) the Orlicz space L_{G_α} where $G_\alpha(t) = \exp(t^\beta) - 1$ with $\beta = 1/\alpha$. We will instead compute $e_n(L_{G_\alpha})$. Suppose $g_1, \ldots, g_n$ are disjoint nonnegative functions with norm one so that

$$\int_0^1 G_\alpha(g_k(t))dt = 1$$

for $k = 1, 2, \ldots, n$. Let $g = \sum_{k=1}^n g_k$. Then

$$\int_0^1 G_\alpha(g(t))dt = n.$$

Now $\|g\|$ is defined by

$$\int_0^1 G_\alpha(\frac{g(t)}{\|g\|})dt = 1.$$

We claim that $\|g\| \leq a_n$ where

$$a_n = \left(\frac{2 \log 8n}{\log(3/2)} \right)^\alpha.$$

We will show that $\int_0^1 G_\alpha(a_n^{-1} g(t))dt \leq 1$. Let us put $E = \{t : g(t)^\beta > 2 \log 8n\}$. Then, for $t \in E$, we have that $\frac{1}{2}(\log 8n)g(t)^\beta > g(t)^\beta$. Thus, again for $t \in E$,

$$(\log 8n)g(t)^\beta > g(t)^\beta + (\log 8n)^2$$

N. J. KALTON

or

$$\frac{g(t)^\beta}{\log 8n} < g(t)^\beta - \log 8n.$$

It follows that

$$\int_E G_\alpha(a_n^{-1}g(t))dt \leq \int_E \exp(a_n^{-\beta}g(t)^\beta)dt$$

$$\leq \int_E \exp\left(\frac{g(t)^\beta}{\log 8n}\right) dt$$

$$\leq \frac{1}{8n} \int_E \exp(g(t)^\beta)dt$$

$$\leq \frac{1}{8n} \int_0^1 (1 + G_\alpha(g(t))dt$$

$$\leq \frac{1}{4}.$$

Let F be the complement of E. Then

$$\int_F G_\alpha(a_n^{-1}g(t))dt \leq \int_0^1 G_\alpha((\log 3/2)^\alpha)dt$$

$$\leq \frac{1}{2}.$$

Combining these statements gives the desired conclusion.■

THEOREM 9.8. *For* $0 < \alpha < 1/2$, *the Orlicz spaces* L_{F_α} *have unique structure as nonatomic Banach lattices.*

We also note here that the preduals of these spaces which coincide with the closure of the simple functions H_{G_α} in L_{G_α} also have unique structure as nonatomic order-continuous Banach lattices (for $\alpha < 1/2$). In the case $\alpha = 1/4$, this example is discussed in Example 2.9 of [6].

10. Isomorphic embeddings

We now turn the problem of characterizing Banach lattices which embed into a given Banach lattice, but without assuming complementation. We shall suppose that X is a good order-continuous Köthe function space on (Δ, λ) and that Y is a good order-continuous Köthe function space on a probability space (K, μ). Let $A : X \to Y$ be an isomorphic embedding; we suppose that M is a constant so that:

$$M^{-1}\|f\|_X \leq \|Af\|_Y \leq M\|f\|_X.$$

We now follow the ideas presented in Section 4. Define $Q_n : CS_n \to L_0(\mu)$ to be linear and satisfy $Q_n\chi_E = |Ah_E|^2$ for $E \in \mathcal{A}_n$.

We define a Haar system in X to be a sequence $(\phi_n)_{n=0}^\infty$ of the form $\phi_0 = \chi_\Delta$ and $\phi_n = \chi_{F_{2n}} - \chi_{F_{2n+1}}$ for $n \geq 1$ where (F_n) is a sequence of clopen sets with $F_1 = \Delta$ and $F_n = F_{2n} \cup F_{2n+1}$ with $\lambda(F_{2n}) = \lambda(F_{2n+1}) = \frac{1}{2}\lambda(F_n)$. The closed linear span $[\phi_n]$ is then a sublattice of the form $X(\Sigma)$ for a certain sub-σ-algebra of the Borel sets, on which λ is non-atomic.

LEMMA 10.1. *Suppose* $\lim_{k\to\infty} Q_k\chi_\Delta = 0$ *in* $L_0(\mu)$. *Then there is a Haar system in* X *spanning a sublattice* $X(\Sigma)$ *which is an unconditional basis for* $X(\Sigma)$ *and is equivalent to a disjoint sequence in* Y.

PROOF: We construct, by induction, a sequence $(F_n)_{n=1}^\infty$ of clopen subsets of Δ and a sequence H_n of Borel subsets of K so that $F_1 = \Delta$, for each n, $F_n = F_{2n} \cup F_{2n+1}$ and $\lambda(F_{2n}) = \lambda(F_{2n+1}) = \frac{1}{2}\lambda(F_n)$, and if $\phi_0 = \chi_\Delta$ and $\phi_n = \chi_{F_{2n}} - \chi_{F_{2n+1}}$ for $n \geq 1$ then, for $n \geq 1$,

$$(\chi_K - \chi_{H_n})|A\phi_n| \leq 2^{-(n+4)} M^{-1} n^{-1} (\chi_K - \chi_{H_n})$$

and

$$\|A\phi_k\chi_{H_n}\|_Y \leq 2^{-(n+4)} k^{-1} M^{-1}$$

for $1 \leq k \leq n-1$.

Suppose $F_1, \ldots, F_{2n-1}$ and $H_1, \ldots, H_{n-1}$ have been selected. Then $F_n \in \mathcal{C}_m$ for some suitable large m. By the order-continuity of Y we may find $\eta = \eta_n > 0$ so that if $\mu E < \eta$ then for $1 \leq j \leq n-1$

$$\|A\phi_j \chi_E\|_Y < 2^{-(n+4)} j^{-1} M^{-1}.$$

Now as $k \to \infty$ we have that $Q_k \chi_{F_n}$ is bounded in $L_{1/2}$ and $Q_k \chi_{F_n} \to 0$ in measure. Hence $\lim_{k\to\infty} \|Q_k \chi_{F_n}\|_{1/4} = 0$. Now for $k \geq m$,

$$Q_k \chi_{F_n} = \sum_{E \in \mathcal{A}_k, E \subset F_n} |Ah_E|^2$$

and so

$$\lim_{k\to\infty} \int_K \Big(\sum_{E \in \mathcal{A}_k, E \subset F_n} |Ah_E|^2 \Big)^{1/4} d\mu = 0.$$

It then follows from Khintchine's inequality that

$$\lim_{k\to\infty} \operatorname*{Ave}_{\epsilon_E = \pm 1} \int \Big| \sum_{\substack{E \in \mathcal{A}_k \\ E \subset F_n}} \epsilon_E Ah_E \Big|^{1/2} d\mu = 0.$$

Thus, for large enough $k \geq m$, there is a choice of signs $\epsilon_E = \pm 1$ such that

$$\int_K \Big| \sum_{\substack{E \in \mathcal{A}_k \\ E \subset F_n}} \epsilon_E Ah_E \Big|^{1/2} d\mu < \sqrt{\frac{\eta^2}{2^{n+4} Mn}}.$$

At this point we define F_{2n}, F_{2n+1} to be disjoint clopen subsets of F_n such that

$$\sum \epsilon_E h_E = \chi_{F_{2n}} - \chi_{F_{2n+1}} = \phi_n.$$

Now let $H_n = \{|A\phi_n| > 2^{-(n+4)} M^{-1} n^{-1}\}$. Then $\mu H_n < \eta$. Thus for $1 \leq j \leq n-1$, we have $\|A\phi_j \chi_{H_n}\|_Y \leq 2^{-(n+4)} j^{-1} M^{-1}$. This completes the inductive construction.

Next let $G_n = H_n \setminus \cup_{k=n+1}^\infty H_k$ for $n \geq 1$. Then

$$\|A\phi_n(\chi_{H_n} - \chi_{G_n})\|_Y \leq \sum_{k=n+1}^\infty 2^{-(k+4)} M^{-1} n^{-1} \leq 2^{-(n+4)} M^{-1} n^{-1}.$$

Also

$$\|A\phi_n(\chi_K - \chi_{H_n})\|_Y \leq 2^{-(n+4)} M^{-1} n^{-1} \|\chi_K\|_Y \leq 2^{-(n+3)} M^{-1} n^{-1}.$$

Combining,

$$\|A\phi_n - (A\phi_n)\chi_{G_n}\|_Y \leq 2^{-(n+2)}M^{-1}n^{-1}.$$

On the other hand,

$$\|A\phi_n\|_Y \geq M^{-1}\|\phi_n\|_X \geq M^{-1}\|\phi_n\|_1 \geq M^{-1}n^{-1}.$$

Hence

$$\|A\phi_n\chi_{G_n}\|_Y \geq (2Mn)^{-1}$$

for all $n \geq 1$. Thus

$$\sum_{n=1}^{\infty}\|A\phi_n - (A\phi_n)\chi_{G_n}\|_Y\|A\phi_n\chi_{G_n}\|_Y^{-1} < 1$$

and as $((A\phi_n)\chi_{G_n})_{n=1}^{\infty}$ is a 1-unconditional basic sequence it follows that $(\phi_n)_{n=0}^{\infty}$ is an unconditional basis for its closed linear span $X(\Sigma)$. ∎

LEMMA 10.2. *If the Haar functions form an unconditional basis of X then ℓ_2 is isomorphic to a complemented subspace of X.*

PROOF: By Theorem 1.d.6 of [33] (and the argument therein) the Rademacher functions $r_n = \sum_{E\in\mathcal{A}_n} h_E$ form an unconditional basic sequence equivalent to the ℓ_2-basis in both X and X^*. It follows quickly that the map

$$Pf = \sum_{n=1}^{\infty}(\int fr_n d\lambda)r_n$$

defines a bounded projection onto $[r_n]$. ∎

LEMMA 10.3. *One of the following three alternatives holds:*
(a) There is a Haar system in X spanning a sublattice $X(\Sigma)$ which is an unconditional basis for $X(\Sigma)$ and is equivalent to a disjoint sequence in Y.
(b) There is a sub-$\sigma-$algebra Σ of $\mathcal{B}(\Delta)$ and a constant $c > 0$, so that for $f \in X(\Sigma)$ we have $\|f\|_X \geq c\|f\|_2$.
(c) There exists a non-trivial lattice homomorphism of X into Y.

PROOF: We will assume that (a) and (c) fail. Thus by Lemma 10.1 we do not have $\lim_{k\to\infty} Q_k\chi_{\Delta} = 0$, in $L_0(\mu)$. We pass to a subsequence n_k so that that there exists

$\delta > 0$ and A_k Borel in K with $\mu A_k \geq \delta$ and $Q_{n_k}(\chi_\Delta) > \delta \chi_{A_k}$. By passing to a further subsequence we can assume that

$$Qf = \lim_{N \to \infty} \frac{1}{N} \sum_{k=1}^N Q_{n_k} f$$

exists in $L_0(\mu)$ for all $f \in CS(\Delta)$, where as usual $Q_n f = 0$ when $f \notin CS_n(\Delta)$. Arguing as in Lemma 4.3 we conclude that

$$\|Qf\|_{Y,1/2} \leq K_G^2 M^2 \|f\|_{X,1/2}$$

and hence that Q extends to a positive measure-continuous operator $Q : X_{1/2} \to Y_{1/2,max}$.

We next argue that if (c) fails then ν_s^Q is almost everywhere a continuous measure. Indeed if not then as in Section 6, we can find $a \in L_\infty(\mu)$ with $a \geq 0$ and $a > 0$ on a set of positive measure and a Borel map $\sigma : K \to \Delta$ so that $\nu_s^Q \geq a(s)\delta_{\sigma s}$ a.e. Then if we set $Tf = (a(s))^{1/2} f(\sigma s)$ then T is a nontrivial lattice homomorphism of X into Y.

Thus we may assume that ν_s^Q is continuous almost everywhere. We now essentially duplicate the arguments of Lemma 6.2 and Theorem 6.3. We select H of positive measure in K so that $Q\chi_\Delta$ is bounded on H and then set $Q_0 f = \chi_H Qf$ for $f \in X$. We shall define a Borel measure on K by $\gamma(F) = \int_K (Q_0 \chi_\Delta)^{1/2} d\mu$. By assumption we have $\gamma(K) > 0$. Let $(\eta_n)_{n=0}^\infty$ be a sequence of positive reals such that $\sum_{n=0}^\infty 2^n \eta_n < \frac{1}{2} \min(1, \gamma(K))$.

Let $G(\emptyset) = \Delta$. Then for each finite sequence $(\epsilon_1, \ldots, \epsilon_n)$ with $\epsilon_k = \pm 1$ we can define a clopen subset $G(\epsilon_1, \ldots, \epsilon_n)$ of Δ and a Borel subset $F(\epsilon_1, \ldots, \epsilon_n)$ of K_2 so that

$$G(\epsilon_1, \ldots, \epsilon_n) = G(\epsilon_1, \ldots, \epsilon_n, 1) \cup G(\epsilon_1, \ldots, \epsilon_n, -1)$$

$$\lambda(G(\epsilon_1, \ldots, \epsilon_n)) = 2^{-n}$$

$$\gamma(K \setminus F(\epsilon_1, \ldots, \epsilon_n)) < \eta_n$$

$$\chi_{F(\epsilon_1, \ldots, \epsilon_n)} |Q_0(\chi_{G(\epsilon_1, \ldots, \epsilon_n, 1)} - \frac{1}{2}\chi_{G(\epsilon_1, \ldots, \epsilon_n)})| \leq \eta_n Q_0 \chi_{G(\epsilon_1, \ldots, \epsilon_n)}$$

It is clear that this can be done by repeated application of Lemma 6.2 for suitably large D (in this case $u_j = 0$).

We now let $F = \cap_{n=1}^\infty \cap_{\epsilon_k = \pm 1} F(\epsilon_1, \ldots, \epsilon_n)$, so that $\gamma(K \setminus F) \leq \sum 2^n \eta_n < \frac{1}{2}\gamma(K)$. Thus $\gamma(F) > \frac{1}{2}\gamma(K) > 0$. Further

$$\chi_F |Q_0 \chi_{G(\epsilon_1, \ldots, \epsilon_n)} - \frac{1}{2^n} T\chi_\Delta| \leq (\sum_{k=0}^{n-1} 2^k \eta_k) 2^{-n} Q_0 \chi_\Delta$$

$$\leq 2^{-(n+1)} Q_0 \chi_\Delta$$

so that

$$\chi_F Q_0(\chi_{G(\epsilon_1,\dots,\epsilon_n)}) \geq \frac{1}{2} 2^{-n} \chi_F Q_0 \chi_\Delta.$$

Now we conclude that for any f of the form,

$$f = \sum_{\epsilon_i = \pm 1} a(\epsilon_1, \dots, \epsilon_n) \chi_{G(\epsilon_1,\dots,\epsilon_n)},$$

we have

$$\frac{1}{2} \left(\int f^2 d\lambda \right) \chi_F Q_0 \chi_\Delta \leq Q_0(f^2)$$

and hence

$$\frac{1}{2} \int f^2 d\lambda (\gamma(F))^2 \leq \|Q_0(f^2)\|_{1/2} \leq \|Q_0(f^2)\|_{Y,1/2}.$$

Hence for some suitable $c > 0$,

$$c \int f^2 d\lambda \leq \|f^2\|_{X,1/2} = \|f\|_X.$$

This clearly extends to the full σ-algebra Σ generated by all $G(\epsilon_1, \dots, \epsilon_n)$, $n \in \mathbf{N}$.∎

We now combine these results to give some applications:

THEOREM 10.4. *Suppose X and Y are separable order-continuous nonatomic Banach lattices such that X embeds into Y. Suppose further that either (1) X and Y are both strictly 2-concave or (2) $\liminf d_n(X)(\log n)^{-1/2} = 0$. Then there is a nontrivial lattice homomorphism of X into Y.*

PROOF: We can suppose that X and Y are Köthe function spaces as above. We simply show that in cases (1) and (2) both (a) and (b) of Lemma 10.3 are excluded. Suppose first we have (1). Then if (a) holds, Y contains a sublattice lattice-isomorphic to ℓ_2 and cannot be strictly 2-concave. If (b) holds then for $f \in X(\Sigma)$ we can find for each N disjoint sets $E_k; 1 \leq k \leq n$, in Σ such that $\|f\chi_{E_k}\|_2 = N^{-1/2}\|f\|_2$. Hence if X is q-concave for some $a > 0$,

$$\|f\|_X \geq a \left(\sum \|f\chi_{E_k}\|^q \right)^{1/q} \geq ac\|f\|_2 N^{1/q - 1/2}$$

which again leads to a contradiction. Now suppose (2) holds; then (a) is excluded because no sublattice of X can contain a complemented Hilbertian subspace by Theorem 9.4. On the other hand X is strictly 2-concave (Lemma 9.3) so that the above reasoning excludes (b).∎

Now we consider the case when X is an r.i. space on $[0,1]$. The following result strengthens Theorem 5.1 of [22] (see also Theorem 6.1 of [22]).

THEOREM 10.5. *Let X be a separable order-continuous r.i. space on $[0,1]$ and let Y be an order-continuous Banach lattice. If X embeds into Y then one of the following three (non-exclusive) possibilities holds:*

(a) The Haar system is an unconditional basis for X and Y contains a sublattice lattice-isomorphic to the Haar represntation of X.

(b) There exists a constant $c > 0$ so that for all $f \in X$, $\|f\|_X \geq c\|f\|_2$,

(c) There is a nontrivial lattice homomorphism of X into Y.

If further Y is an r.i. space on $[0,1]$ or $[0,\infty)$ then (c) is equivalent to:

(d) There exists $c > 0$ so that for all $f \in X$, $\|f\|_X \geq c\|f\|_Y$.

PROOF: The first part of the assertion is immediate from Lemma 10.3. For the equivalence of (c) and (c'), let us suppose that Y is an r.i. space on $[0,\infty)$ and that $V : X \to Y$ is a nontrivial lattice homomorphism. Thus V is of the form

$$Vf(s) = a(s)f(\sigma s)$$

where $a \in L_0[0,\infty)$ is nonnegative and is strictly positive on a set of positive measure, and $\sigma : [0,\infty) \to [0,1]$ is a Borel map. We may clearly suppose that $a = \chi_E$ where E is a Borel set of measure less than one. Define a measure ν on $[0,1]$ by $\nu F = \lambda(\sigma^{-1}F \cap E)$. Then it easily follows that ν is $\lambda-$continuous on $[0,1]$ and is nontrivial. Hence there exists a Borel set G of positive measure so that for suitable $0 < \alpha < \beta < \infty$ we have $\alpha\lambda(H) \leq \nu(H) \leq \beta\lambda(H)$ whenever H is a Borel subset of G. Now if f is supported on G then

$$\|D_\alpha f\|_Y \leq \|Vf\|_Y \leq C\|f\|_X$$

for a suitable constant C and hence (d) follows.∎

We conclude by considering the special case when $X = L_p[0,1]$. The following result is a generalization of Lemma 7.13 of [5].

LEMMA 10.6. *Suppose $1 \leq p < \infty$ and that Y is a separable order-continuous p-convex Banach lattice. Suppose that there is a nontrivial lattice homomorphism $V : L_p \to Y$. Then Y contains a complemented sublattice Z which is lattice-isomorphic to L_p.*

PROOF: We can suppose that Y is a good order-continous Köthe function space on a probability space (K, μ). We can suppose that $Vf = \chi_E f \circ \sigma$ where $\mu E > 0$ and $\sigma : K \to$

$[0, 1]$ is a Borel map. Consider the Köthe function space $Z = Y_{1/p}$. Then there exists $\phi \in Z_+^*$ so that $\phi > 0$ a.e. and $\int \phi \chi_E d\mu > 0$. Thus for $f \in Y$,

$$\int |f|^p \phi d\mu \leq \||f|^p\|_Z = \|f\|_Y^p.$$

Define ν to be measure on $[0, 1]$ given by

$$\nu(F) = \int_{\sigma^{-1} F \cap E} \phi \, d\mu.$$

Then it is again to easy to see that ν is λ−continuous and nonzero. Thus there is Borel subset G of $[0, 1]$ of positive measure such that for some $\alpha > 0$ and all Borel subsets H of G we have $\nu(H) \geq \alpha \lambda(H)$. Now if $f \in L_p(G)$,

$$\int |Vf|^p \phi d\mu = \int_E |f \circ \sigma|^p \phi d\mu$$
$$= \int_G |f|^p d\nu$$
$$\geq \alpha \|f\|_p^p.$$

It follows that V maps $L_p(G)$ isomorphically onto a sublattice L of $L_p(\phi d\mu)$. Let P be a norm one projection of $L_p(\phi d\mu)$ onto L. Since on L the Y and $L_p(\phi d\mu)$ norms are equivalent P also induces a projection of Y onto L.■

The next theorem is an improvement (in the case $p = 1$) and a generalization of the result of [25].

THEOREM 10.7. Suppose $1 \leq p < 2$ and suppose that Y is a separable order-continuous p-convex Banach lattice. Suppose L_p embeds into Y. Then either:

(a) $p > 1$ and Y contains a sublattice lattice-isomorphic to the Haar representation of L_p, or

(b) Y contains a complemented sublattice lattice-isomorphic to L_p.

PROOF: This is now immediate from Theorem 10.5 and Lemma 10.6.■

Obviously if we assume that Y is strictly 2-concave then case (a) is excluded and we must have that L_p is lattice-isomorphic to a complemented sublattice of Y. However, if we do not require complementation the same conclusion can be obtained in Y is merely of type p (in the case $p > 1$.) In order to prove this we use a minor modification of a result from [26].

PROPOSITION 10.8. *Suppose $0 < p < 1$ and that Z is a p-Köthe space on some Polish probability space (K, μ), which is r-concave for some $r < \infty$. Suppose there exists a positive operator $V : L_p \to Z$ such that for some $c > 0$ and all nonnegative $f \in L_p$ we have $\|Vf\|_Z \geq c\|f\|_p$. Then there is an embedding of L_p into Z which is a lattice isomorphism.*

PROOF: This is essentially proved in [26], Theorem 8.7. The hypotheses there are that V is an embedding (not necessarily positive). Nevertheless the result quoted is proved by exactly the same reasoning; one need only observe that it suffices to consider operators satisfying a lower estimate on the positive cone in place of embeddings.

THEOREM 10.9. *Suppose $1 < p < 2$. Let Y be a separable strictly 2-concave Banach lattice of type $p > 1$. Suppose that L_p embeds into Y. Then Y contains a sublattice lattice-isomorphic to L_p.*

PROOF: We shall suppose that $L_p = L_p(\Delta, \lambda)$ and Y is a good Köthe function space on the probability space (K, μ). We define $Z = Y_{1/2}$. Let us observe that Z can be equivalently normed as a $p/2$–Köthe function space. Indeed if $g_i \in Z_+$ for $1 \leq i \leq n$ then $g_i = f_i^2$ where $f_i \in Y$. Then for constants C, C' depending only on Y we have:

$$\|\sum_{i=1}^{n} g_i\|_Z = \|(\sum_{i=1}^{n} f_i^2)^{1/2}\|_Y^2$$

$$\leq C \mathrm{Ave}_{\epsilon_i = \pm 1} \|\sum_{i=1}^{n} \epsilon_i f_i\|_Y^2$$

$$\leq CC' (\sum_{i=1}^{n} \|f_i\|^p)^{2/p}$$

$$\leq CC' (\sum_{i=1}^{n} \|g_i\|^{p/2})^{2/p}.$$

Thus Z is a $p/2$–Köthe function space. Further since Y is q-concave for some $q < 2$, Z is $r = q/2$–concave where $r < 1$.

Now as in the discussion before Lemma 10.1 we suppose $A : L_p \to Y$ is an embedding satisfying $M^{-1}\|f\|_p \leq \|Af\|_Y \leq M\|f\|_p$ for all $f \in L_p(\Delta, \lambda)$. Define $Q_n : CS_n \to L_0(\mu)$ as before. As in Lemma 10.3 we may select, by Komlos's theorem, a subsequence so that for every $f \in CS(\Delta)$

$$Qf = \lim_{N \to \infty} \frac{1}{N} \sum_{k=1}^{N} Q_{n_k} f$$

exists $\mu - a.e.$ From Lemma 4.3 we conclude that Q is extends to a positive bounded operator from $L_{p/2}$ to Z. It remains to show that Q satisfying a lower estimate for positive functions.

Suppose $f \in CS_{n_l}$ is nonnegative and $\|f\|_{p/2} = 1$. Let $Qf = h \in Z_+$. Then we may find a sequence N_j tending to infinity so that if

$$h_j = \frac{1}{N_j} \sum_{k=l}^{N_j} Q_{n_k} f$$

then we can write, by a standard gliding hump technique,

$$h_j = h + \phi_j + \psi_j$$

where $(\phi_j)_{j=1}^{\infty}$ is a sequence of disjointly supported functions with $0 \leq \phi_j \leq h_j$ and $\|\psi_j\|_Z \to 0$. We will argue that $\|\phi_j\|_Z \to 0$ as well. If not we may pass to a subsequence so that $\|\phi_j\|_Z \geq \delta > 0$ for all j. Then for all m and a suitable constant $a > 0$ since Z is r-concave,

$$cm^{1/r-1}\delta \leq \|\frac{1}{m} \sum_{j=1}^{m} \phi_j\|_Z$$

$$\leq \|\frac{1}{m} \sum_{j=1}^{m} h_j\|_Z$$

$$= \|\frac{1}{m} \sum_{j=1}^{m} \frac{1}{N_j} \sum_{k=l}^{N_j} Q_{n_k} f\|_Z$$

$$\leq K_G^2 M^2 \|f\|_{p/2}$$

by applying Lemma 4.2. But this yields a contradiction as $m \to \infty$ and so we do indeed have $\lim \|\phi_j\|_Z = 0$. Hence $\lim \|h_j - h\|_Z = 0$.

If we write $f = \sum_{E \in \mathcal{A}_{n_l}} \alpha_E^2 \chi_E$ then

$$h_j = \frac{1}{N_j} \sum_{k=l}^{N_j} \sum_{E \in \mathcal{A}_{n_l}} \sum_{\substack{F \in \mathcal{A}_{n_k} \\ F \subset E}} \alpha_E^2 |Ah_F|^2$$

and hence for suitable constants $c, c' > 0$ since Y has cotype 2,

$$\|h_j\|_Z^{1/2} = \|h_j^{1/2}\|_Y$$

$$\geq c \operatorname*{Ave}_{\epsilon_F = \pm 1} \Big\| \sum_{k=l}^{N_j} \sum_{E \in \mathcal{A}_{n_l}} \sum_{\substack{F \in \mathcal{A}_{n_k} \\ F \subseteq E}} \epsilon_F \frac{\alpha_E}{N_j^{1/2}} A h_F \Big\|_Y$$

$$\geq c' \Big\| \Big(\frac{1}{N_j} \sum_{k=l}^{N_j} \sum_{E \in \mathcal{A}_{n_l}} \sum_{\substack{F \in \mathcal{A}_{n_k} \\ F \subseteq E}} \alpha_E^2 \chi_F \Big)^{1/2} \Big\|_p$$

$$\geq c' \left(\frac{N_j - l}{N_j} \right)^{1/2}.$$

Hence $\liminf_{j \to \infty} \|h_j\|_Z \geq (c')^2$ and so $\|Qf\|_Z \geq (c')^2$.

The theorem will now follow from Proposition 10.8.∎

We remark that the Lorentz spaces $L(p, q)$ are of type p when $1 < p \leq q < \infty$. It is in fact known that ℓ_p does not embed into $L(p, q)$ ([10]). See [7], [8], [9] and [10] for general results on subspaces of Lorentz spaces.

References

1. Y. A. Abramovich, On the maximal normed extension of semi-ordered spaces, Vestnik Leningrad Univ. Math. Mech. Astronomy, 1 (1970) 7-17.

2. Y. A. Abramovich and P. Wojtaszczyk, On the uniqueness of order in the spaces ℓ_p and $L_p[0,1]$, Mat. Zametki 18 (1975) 313-325.

3. C.D. Aliprantis and O. Burkinshaw, *Positive operators,* Academic Press, Orlando, 1985.

4. D. Amir and J. Lindenstrauss, The structure of weakly compact sets in Banach spaces, Ann. Math. (2) 88 (1968) 35-46.

5. J. Bourgain, P.G. Casazza, J. Lindenstrauss and L. Tzafriri, *Banach spaces with a unique unconditional basis, up to permutation,* Mem. Amer. Math. Soc. No. 322, 1985.

6. J. Bourgain, N.J. Kalton and L. Tzafriri, Geometry of finite dimensional subspaces and quotients of L_p, G. A. F. A. Seminar (J. Lindentrauss, V. D. Milman, editors) Springer Lecture Notes 1376 (1989) 138-175.

7. N.L. Carothers, Rearrangement-invariant subspaces of Lorentz function spaces, Israel J. Math. 40 (1981) 217-228.

8. N.L. Carothers, Rearrangement-invariant subspaces of Lorentz function spaces II, Rocky Mountain J. Math. 17 (1987) 607-616.

9. N.L. Carothers and S.J. Dilworth, Geometry of Lorentz spaces via interpolation, Longhorn Notes, University of Texas, 1985-6, 107-134.

10. N.L. Carothers and S.J. Dilworth, Subspaces of $L_{p,q}$, Proc. Amer. Math. Soc. 104 (1988) 537-545.

11. N.L. Carothers and S.J. Dilworth, Some Banach space embeddings of classical function spaces, to appear.

12. P.G. Casazza, N.J. Kalton and L. Tzafriri, Decompositions of Banach lattices into direct sums, Trans. Amer. Math. Soc. 304 (1987) 771-800.

13. P.G. Casazza, N.J. Kalton and L. Tzafriri, Uniqueness of uncondtional and symmetric strutures in finite dimensional spaces, Illinois J. Math. 34 (1990) 793-836.

14. D.L. Cohn, *Measure Theory*, Birkhauser, Boston 1980.

15. S.J. Dilworth, Intersection of Lebesgue spaces L_1 and L_2, Proc. Amer. Math. Soc. 103 (1988) 1185-1188.

16. S.J. Dilworth, A scale of lincar spaces closely related to the L_p scale, Illinois J. Math. 34 (1990) 140-158.

17. L.E. Dor and T. Starbird, Projections of L_p onto subspaces spanned by independent random variables, Comp. Math. 39 (1979) 141-175.

18. I. Edelstein and P. Wojtaszczyk, On projections and unconditional bases in direct sums of Banach spaces, Studia Math. 56 (1976) 263-276.

19. T.A. Gillespie, Factorization in Banach function spaces, Indag. Math. 43 (1981) 287-300.

20. W.T. Gowers, A finite-dimensional Banach space with two non-equivalent symmetric bases, to appear.

21. F.L. Hernandez and B. Rodriguez-Salinas, On ℓ^p−complemented copies in Orlicz spaces, Israel J. Math. 62 (1988) 37-55.

22. W.B. Johnson, B. Maurey, G. Schechtman and L. Tzafriri, *Symmetric structures in Banach spaces*, Mem. Amer. Math. Soc. No. 217, 1979.

23. W. B. Johnson and G. Schechtman, Sums of independent random variables in r.i. function spaces, Ann. Prob. 17 (1989) 789-800.

24. N.J. Kalton, The endomorphisms of L_p, $0 \leq p \leq 1$, Indiana Univ. Math. J. 27 (1978) 353-381.

25. N.J. Kalton, Embedding L_1 in a Banach lattice, Israel J. Math. 32 (1979) 209-220.

26. N.J. Kalton, Linear operators on L_p for $0 < p < 1$, Trans. Amer. Math. Soc. 259 (1980) 319-355.

27. N.J. Kalton, Banach spaces embedding into L_0, Israel J. Math. 52 (1985) 305-319.

28. J. Komlos, A generalization of a problem of Steinhaus, Acta. Math. Acad. Sci. Hungar. 18 (1967) 217-229.

29. J.L. Krivine, Théorèmes de factorisation dans les espaces reticules, Seminaire Maurey-Schwartz 1973-74, Exposes 22-23, Ecole Polytechnique, Paris.

30. J. Lindenstrauss and A. Pelczynski, Absolutely summing operators in $\mathcal{L}_p$–spaces and their applications, Studia Math. 29 (1968) 275-326.

31. J. Lindenstrauss and L. Tzafriri, On Orlicz sequence spaces, Israel J. Math. 10 (1971) 379-390.

32. J. Lindenstrauss and L. Tzafriri, *Classical Banach spaces I, Sequence spaces,* Springer Verlag, Berlin, Heidelberg, New York 1977.

33. J. Lindenstrauss and L. Tzafriri, *Classical Banach spaces II, Function spaces,* Springer Verlag, Berlin, Heidelberg, New York 1979.

34. J. Lindenstrauss and M. Zippin, Banach spaces with a unique unconditional basis, J. Functional Analysis 3 (1969) 115-125.

35. G. Y. Lozanovskii, On topologically reflexive KB-spaces, Dokl. Akad. Nauk. USSR 158 (1964) 516-519.

36. G.Y. Lozanovskii, On some Banach lattices, Siberian Math. J. 10 (1969) 419-430.

37. B. Maurey, Type et cotype dans les espaces munis de structures locales inconditionelles, Seminaire Maurey-Schwartz 1973-74 Expose 24-25, Ecole Polytechnique, Paris.

38. E.M. Nikishin, Resonance theorems and superlinear operators, Uspehi Mat. Nauk. 25 (1970) 129-191 (Russian Math. Surveys 25 (1970) 124-187.)

39. E.M. Nikishin, On systems of convergence in measure to ℓ_2, Mat. Zametki 13 (1973) 337-340 (Math. Notes 13 (1973) 205-207.)

40. Y. Raynaud, Sous espaces ℓ_r et geometrie des espaces $L_p(L_q)$ et L_ϕ, C. R. Acad. Sci. (Paris) Series I, 301 (1985) 299-302.

41. Y. Raynaud and C. Schutt, Some results on symmetric subspaces of L_1, Studia Math. 89 (1988) 27-35.

42. C.J. Read, A Banach space with, up to equivalence, precisely two symmetric bases, Israel J. Math. 40 (1981) 33-53.

43. H.H. Schaefer, *Banach lattices and positive operators,* Springer Verlag, Berlin, Heidelberg, New York 1974.

44. C. Schutt, On the uniqueness of symmetric bases in finite-dimensional Banach spaces, Israel J. Math. 40 (1981) 97-117.

45. C. Schutt, Lorentz spaces that are isomorphic to subspaces of L_1, Trans. Amer. Math.

Soc. 314 (1989) 583-595.

46. A.R. Sourour, Pseudo-integral operators, Trans. Amer. Math. Soc. 253 (1979) 339-363.

47. L. Weis, On the representation of positive operators by random measures, Trans. Amer. Math. Soc. 285 (1984) 535-564.

48. P. Wojtaszczyk, On complemented subspaces and unconditional bases in $\ell_p \oplus \ell_q$, Studia Math. 47 (1973) 197-206.

49. P. Wojtaszczyk, On projections and unconditional bases in direct sums of Banach spaces II, Studia Math. 62 (1978) 193-201.

50. P. Wojtaszczyk, *Banach spaces for analysts,* Cambridge University Press, 1991.

N.J. Kalton

Department of Mathematics, University of Missouri

Columbia, Mo. 65211

Editorial Information

To be published in the *Memoirs*, a paper must be correct, new, nontrivial, and significant. Further, it must be well written and of interest to a substantial number of mathematicians. Piecemeal results, such as an inconclusive step toward an unproved major theorem or a minor variation on a known result, are in general not acceptable for publication. *Transactions* Editors shall solicit and encourage publication of worthy papers. Papers appearing in *Memoirs* are generally longer than those appearing in *Transactions* with which it shares an editorial committee.

As of March 1, 1993, the backlog for this journal was approximately 7 volumes. This estimate is the result of dividing the number of manuscripts for this journal in the Providence office that have not yet gone to the printer on the above date by the average number of monographs per volume over the previous twelve months, reduced by the number of issues published in four months (the time necessary for preparing an issue for the printer). (There are 6 volumes per year, each containing at least 4 numbers.)

A Copyright Transfer Agreement is required before a paper will be published in this journal. By submitting a paper to this journal, authors certify that the manuscript has not been submitted to nor is it under consideration for publication by another journal, conference proceedings, or similar publication.

Information for Authors

Memoirs are printed by photo-offset from camera copy fully prepared by the author. This means that the finished book will look exactly like the copy submitted.

The paper must contain a *descriptive title* and an *abstract* that summarizes the article in language suitable for workers in the general field (algebra, analysis, etc.). The *descriptive title* should be short, but informative; useless or vague phrases such as "some remarks about" or "concerning" should be avoided. The *abstract* should be at least one complete sentence, and at most 300 words. Included with the footnotes to the paper, there should be the 1991 *Mathematics Subject Classification* representing the primary and secondary subjects of the article. This may be followed by a list of *key words and phrases* describing the subject matter of the article and taken from it. A list of the numbers may be found in the annual index of *Mathematical Reviews*, published with the December issue starting in 1990, as well as from the electronic service e-MATH [**telnet e-MATH.ams.org** (or **telnet 130.44.1.100**). Login and password are **e-math**]. For journal abbreviations used in bibliographies, see the list of serials in the latest *Mathematical Reviews* annual index. When the manuscript is submitted, authors should supply the editor with electronic addresses if available. These will be printed after the postal address at the end of each article.

Electronically-prepared manuscripts. The AMS encourages submission of electronically-prepared manuscripts in $\mathcal{A}_{\mathcal{M}}\mathcal{S}$-TEX or $\mathcal{A}_{\mathcal{M}}\mathcal{S}$-LATEX. To this end, the Society has prepared "preprint" style files, specifically the amsppt style of $\mathcal{A}_{\mathcal{M}}\mathcal{S}$-TEX and the amsart style of $\mathcal{A}_{\mathcal{M}}\mathcal{S}$-LATEX, which will simplify the work of authors and of the production staff. Those authors who make use of these style files from the beginning of the writing process will further reduce their own effort.

Guidelines for Preparing Electronic Manuscripts provide additional assistance and are available for use with either $\mathcal{A}_{\mathcal{M}}\mathcal{S}$-TEX or $\mathcal{A}_{\mathcal{M}}\mathcal{S}$-LATEX. Authors with FTP access may obtain these *Guidelines* from the Society's Internet node e-MATH.ams.org (130.44.1.100). For those without FTP access they can be obtained free of charge from the e-mail address guide-elec@math.ams.org (Internet) or from the Publications Department, P. O. Box 6248, Providence, RI 02940-6248. When requesting *Guidelines* please specify which version you want.

Electronic manuscripts should be sent to the Providence office only after the paper has been accepted for publication. Please send electronically prepared manuscript files via e-mail to pub-submit@math.ams.org (Internet) or on diskettes to the Publications Department address listed above. When submitting electronic manuscripts please be sure to include a message indicating in which publication the paper has been accepted.

For papers not prepared electronically, model paper may be obtained free of charge from the Editorial Department at the address below.

Two copies of the paper should be sent directly to the appropriate Editor and the author should keep one copy. At that time authors should indicate if the paper has been prepared using $\mathcal{A}_{\mathcal{M}}\mathcal{S}$-TEX or $\mathcal{A}_{\mathcal{M}}\mathcal{S}$-LATEX. The *Guide for Authors of Memoirs* gives detailed information on preparing papers for *Memoirs* and may be obtained free of charge from AMS, Editorial Department, P. O. Box 6248, Providence, RI 02940-6248. The *Manual for Authors of Mathematical Papers* should be consulted for symbols and style conventions. The *Manual* may be obtained free of charge from the e-mail address cust-serv@math.ams.org or from the Customer Services Department, at the address above.

Any inquiries concerning a paper that has been accepted for publication should be sent directly to the Editorial Department, American Mathematical Society, P. O. Box 6248, Providence, RI 02940-6248.

Recent Titles in This Series

(*Continued from the front of this publication*)

(See the AMS catalog for earlier titles)